Human cytogenetics

a practical approach

TITLES PUBLISHED IN
THE
PRACTICAL APPROACH
SERIES

Human cytogenetics

a practical approach

Edited by

D E Rooney

Cytogenetics Unit, St. Mary's Hospital Medical School,
London, UK

B H Czepulkowski

Department of Medical Oncology, St. Bartholomew's Hospital,
London, UK

 IRL PRESS

Oxford · Washington DC

IRL Press Ltd
P.O. Box 1
Eynsham
Oxford OX8 1JJ
England

British Library Cataloguing in Publication Data

Human cytogenetics : a practical approach.—(Practical
  approach series)
  1. Human genetics 2. Cytogenetics
  I. Rooney, D.E. II. Czepulkowski, B.H.
  III. Series
  611′.01816   QA431

ISBN 0-947946-70-5 (softbound)
ISBN 0-947946-71-3 (hardbound)

Cover illustration. The design for the cover was reproduced from ISCN
(1978): An International System for Human Cytogenetic Nomenclature (1978).
Birth Defects: Original Article Series, Vol XIV, No. 8 (The National
Foundation, New York, 1978).

Printed by Information Printing Ltd, Oxford, England.

Preface

The last 18 years have seen the development of almost all of the techniques currently used in human cytogenetics. Many of them are less than 10 years old. Such is the diversity of approaches to the practical aspects of cytogenetics that it has often been accused of owing more to cookery and superstition than to true science. Indeed, one cytogeneticist on learning of the proposed volume insisted that here was a chance to remove all traces of mysticism. In the same week several laboratories were praying for rain because a prolonged dry spell was wreaking havoc with their chromosome spreading...

Chapters 1, 2 and 3 cover all aspects of chromosome preparation from tissues normally dealt with in a routine diagnostic laboratory. Most hospital laboratories will have their own tried and tested methods and will probably use these chapters together with Chapter 4 as a reference back-up. This volume is, however, also intended for undergraduates, postgraduates and trainee cytogeneticists. We have tried, therefore, to include techniques which have proved to be the most adaptable together with enough theoretical background for the student to make the most effective use of them.

Chapters 5 and 6 cover some specialised areas of human cytogenetics. Not all routine diagnostic laboratories deal with malignancy and very few with meiotic studies. Such laboratories do, however, encounter the occasional bone marrow, tumour or testicular biopsy and there has been a recent demand for protocols. These aspects of human cytogenetics together with those covered in Chapter 7 are also of relevance to University and Research Institute based scientists.

There has been a wide and rapid expansion of research into cytogenetics because of the interest in human gene mapping and cancer cytogenetics, hence the inclusion of Chapter 7. It is at this point that our volume merges into another in this series, namely 'Human Genetic Diseases - A Practical Approach' edited by Kay Davies. Two of the topics, *in situ* hybridisation and flow karyotyping are covered in both books, but we have included them here against a cytogenetic as opposed to a genetic background. The majority of the chapter is, however, devoted to the culture of somatic cell hybrids. Chapter 7 is essentially an amalgamation of three 'mini chapters' each one being written independently by the three authors as follows; *in situ* hybridisation (Sue Malcolm), manipulation of somatic cell hybrids for the analysis of the human genome (John Cowell) and flow cytometric analysis of human chromosomes (Bryan Young).

We dedicate this book to Shaun Robinson.

Denise Rooney and Barbara Czepulkowski

Acknowledgements

We wish to thank our authors for the excellent chapters they have contributed. Our thanks are also due to Richard Moreland and Oonagh Heron for help with the proof reading. We are particularly indebted to Dr. Dulcie V. Coleman for her encouragement and support.

Contributors

P.A.Benn
Lifecodes Corporation, 4 Westchester Plaza, Elmsford, NY 10523, USA

J.K.Cowell
Laboratory of Molecular Genetics, Department of Haematology and Oncology, Institute of Child Health, 30 Guilford Street, London WC1, UK

M.R.Creasy
Regional Cytogenetics Laboratory, East Birmingham Hospital, Bordesley Green East, Birmingham B9 5ST, UK

B.H.Czepulkowski
Department of Medical Oncology, St. Bartholomew's Hospital, West Smithfield, London EC1 7BE, UK

C.J.Harrison
Department of Cell Biology, Paterson Laboratories, Christie Hospital, Manchester M20 9BX, UK

M.A.Hultén
Regional Cytogenetics Laboratory, East Birmingham Hospital, Bordesley Green East, Birmingham B9 5ST, UK

J.A.Jonasson
Medical Genetics Department, Churchill Hospital, Headington, Oxford, UK

S.Malcolm
Laboratory of Molecular Genetics, Institute of Child Health, 30 Guilford Street, London WC1, UK

M.A.Perle
Cytogenetics Laboratory, New York University Medical Center, New York, NY 10016, USA

D.E.Rooney
Cytogenetics Unit, St. Mary's Hospital Medical School, Norfolk Place, London W2 1PG, UK

N.Saadallah
Regional Cytogenetics Laboratory, East Birmingham Hospital, Bordesley Green East, Birmingham B9 5ST, UK

G.S.Stephen
Medical Genetics, Department of Genetics, Medical School Buildings, Foresterhill, Aberdeen AB9 2ZD, UK

B.M.N.Wallace
Regional Cytogenetics Laboratory, East Birmingham Hospital, Bordesley Green East, Birmingham B9 5ST, UK

J.L.Watt
Medical Genetics, Department of Genetics, Medical School Buildings, Foresterhill, Aberdeen AB9 2ZD, UK

B.D.Young
Department of Medical Oncology, St. Bartholomew's Hospital, West Smithfield, London EC1 7BE, UK

Contents

Abbreviations

ACC	Association of Clinical Cytogenetics
AchE	acetylcholinesterase
AFP	alpha fetoprotein
ALL	acute lymphoblastic leukaemia
ANLL	acute non-lymphocytic leukaemia
ASG	acetic-saline-Giemsa
BrdU	5-bromo-deoxyuridine
CHO	Chinese hamster ovary
CLL	chronic lymphoblastic leukaemia
CML	chronic myeloid leukaemia
CS	calf serum
CVS	chorionic villus sampling
DAPI	4,6-diamino-2-phenyl-indole
dC	deoxycytidine
DHFR	dihydrofolate reductase
DMs	double minutes
DMSO	dimethyl sulphoxide
EBSS	Earle's balanced salt solution
EBV	Epstein Barr virus
EDTA	ethylenediamine tetraacetic acid
FA	folic acid
FAB	French American British
FCS	fetal calf serum
FITC	fluorescein isothiocyanate
fra(X)	fragile X
FUdR	fluorodeoxyuridine
G-banding	Giemsa banding
HAT	hypoxanthine, aminopterin, thymidine
HBSS	Hanks' balanced salt solution
HPRT	hypoxanthine phosphoribosyltransferase
HSR	homogeneously staining regions
IL-2	interleukin 2
LPS	lipopolysaccharide
MEM	Eagles' minimum essential medium
MTX	methotrexate
NCS	newborn calf serum
NHL	non-Hodgkin's lymphomas
NOR-banding	nucleolar organiser region banding
PBS	phosphate-buffered saline
PBSA	PBS without calcium or magnesium ions
PEG	polyethylene glycol
PH[1] chromosome	Philadelphia chromosome
PHA	phytohaemagglutinin
PTA	phototungstic acid
PWM	pokeweed mitogen
Q-banding	quinacrine banding
R-banding	reverse banding

RBC	red blood cells
RFLP	restriction fragment length polymorphisms
SCE	sister chromatid exchange
SFM	serum free medium
T	thymidine
TG	thyoguanine
TK	thymidine kinase
TPA	12-O-tetradecanoylphorbol-13-acetate
WBC	white blood cells
WCC	white cell counts

CHAPTER 1

Tissue Culture Methods in Human Cytogenetics

D.E.ROONEY and B.H.CZEPULKOWSKI

1. INTRODUCTION

The *in vitro* cultivation of human cells and tissues forms an integral part of the work of every diagnostic cytogenetics laboratory, since it is from cells undergoing mitosis that metaphase chromosome spreads are obtained. Spontaneously dividing cells suitable for direct chromosome preparations are found only in the rapidly proliferating tissues of the body such as the gonads, bone marrow and trophoblast, or in tissues with malignancies. In order to obtain metaphase spreads from other cells or tissues, it is necessary to induce cell division artificially. Where possible, this is carried out on peripheral blood since it is easily obtained, and requires only a 48 − 72 h suspension culture in the presence of a mitogen (Chapter 2). Mitosis can be induced in cells from most other tissues by culturing them on a substrate for periods ranging from a few days to several weeks. Unlike blood cultures, which require a single mitogen such as phytohaemagglutinin (PHA) for stimulation, other cultures require many different growth factors and mitogens which are less well defined.

Whereas blood culture is appropriate for chromosomal investigation of most living patients, from newborn babies to adults, there are occasionally indications for additional investigation of other tissues. The tissue most commonly used is the skin biopsy which may, in rare cases, have a different chromosome complement from the lymphocytes. Thus, it is used when investigation of chromosome mosaicism is necessary. It may also be used if blood cultures from a patient have repeatedly failed due to treatment with drugs or to ailments that affect the number of lymphocytes in the blood.

It is sometimes possible to obtain a post-mortem blood sample suitable for culture if the specimen is taken within an hour of death, but this is rarely done in time. Chromosome studies on post-mortem material are often required for fetuses, neonates or very young babies, and tissues such as skin, muscle and cartilage may survive for up to 3 days after death. Chromosome investigations may also be carried out on cultures initiated from 'products of conception' which will normally be a mixture of fetal sac, embryonic and uterine material expelled by spontaneous abortion, or evacuated surgically by suction.

The various tissue and blood cultures described above are all performed when abnormality has been recognised clinically, and a chromosome investigation is required before a diagnosis can be made. It is recognised that there are certain risk groups, such as women in their late thirties or over, who have an increased chance of conceiving chromosomally abnormal progeny. Thus, a chromosome investigation is required to predict the outcome of a pregnancy, allowing sufficient time for a termination to be

1

performed if this is required on the basis of an abnormal result. Prenatal diagnosis of chromosome abnormalities, open neural tube defects and many inborn errors of metabolism and genetic disorders can be achieved using amniotic fluid. The cells present in the amniotic fluid are derived from several different fetal tissues, and must be cultured for chromosome analysis.

Amniocentesis must, however, be performed well into the second trimester of pregnancy (14 weeks at the earliest) for sufficient fetal cells to be present for culture. This results in late terminations of abnormal pregnancies, which are both difficult and traumatic. For this reason amniocentesis is unacceptable to some patients. The most recent advance in the field of prenatal diagnosis has been in the use of chorionic villus sampling (CVS), which has enabled tests to be done in the first trimester of pregnancy (from 8 to 12 weeks) so that simple suction terminations can be performed if necessary.

This chapter deals with the tissue culture methods commonly used for all the above-mentioned tissues. First it is necessary to consider environmental conditions which constitute a basic requirement of all human tissue cultures.

2. ENVIRONMENTAL CONDITIONS FOR OPTIMUM CELL GROWTH

The environmental conditions necessary for optimum cell growth in tissue cultures are influenced by the following.

(i)   Sterility.
(ii)   Culture vessels and substrate.
(iii)   Medium and additives.
(iv)   Temperature.
(v)   pH.
(vi)   Osmotic pressure.
(vii)   Gas tension.

Each of these points will be considered in detail.

2.1 Sterility

Many microorganisms grow more rapidly than human cells in tissue culture, most of which produce toxins that are harmful to the cultured cells. Microorganisms which grow in cultures without killing them will obscure the stained chromosome spreads, even if attempts have been made to dilute out the contaminants by repeated washing. It is, therefore, essential that all cultures are maintained in a sterile condition.

There are two categories of contaminants which are important in tissue culture, those which are normally present on the body, and those which are present in air or water. There are many ways in which the risk of introducing contaminants into cultures can be minimised, the most important of which is a good aseptic technique on the part of the operator.

2.1.1 Aseptic Technique

There are several ways in which the operator may inadvertently introduce contaminants into a culture, the commonest of which are the following.

(i)   Touching sterile items which may come into direct contact with the specimen, culture medium or inside of culture vessels. Two common examples of this are holding a syringe near to the tip when fastening on a needle and brushing the neck of a culture flask against the hand while screwing on the cap.

(ii)   Coughing, speaking, sneezing or breathing directly over sterile items that come into direct contact with the culture.

(iii) Bending over the sterile items so that contaminants from the hair can fall onto them.

(iv)   Putting sterile items on the bench, and then re-using them (e.g., putting the cap of a tissue culture flask down on the bench surface while other work is in progress, and then replacing it).

(v)   Touching part of a sterile item, which may be in direct contact with the specimen, culture medium or inside of the culture vessel, with a non-sterile surface (e.g., brushing a pipette tip against a reagent bottle, and then using it to withdraw medium from a flask of cells).

(vi)   Moving the hand or arm across an opened culture vessel which may cause contaminants to fall into the culture (e.g., reaching across an open Petri dish to pick up a bottle).

Many laboratories advocate the wearing of sterile gloves, face-masks, and theatre caps while culturing, which considerably reduces the above risks. These items are not strictly necessary as long as good aseptic technique is observed, although the wearing of a clean laboratory coat over street clothes is advisable to reduce the movement of dust. Most laboratories carry out all tissue culture procedures in a safety cabinet which provides a flow of sterile air within which to work. These should ideally have a vertical air-flow to protect the specimen from airborne contamination, and have a vent to the outside so that contaminants from infected samples are not blown towards the operator. Many laboratories, however, stil use the laminar flow cabinet which protects the culture from airborne contaminants but which blows air over the culture and towards the operator thereby exposing him to the risk of infection. The laminar flow cabinet should therefore *never* be used for a specimen known to have a high risk of infection.

Ideally, it should be possible to carry out sterile tissue culture without any of the above conditions, although this is rarely likely to be necessary. It is important that the tissue culture room be kept clean and tidy, and that only the items essential for the work be kept in it. Storage of unnecessary items will encourage the accumulation of dust which will harbour contaminants. It is important that there is as little movement in the room as possible while culturing is in progress so that dust is not unduly disturbed. Thus, thoroughfare should be restricted at these times, and the operator should be allowed to work uninterrupted. It is recommended that as few people as possible should be allowed into the room while culturing is in progress. On no account should a water bath or a sink be situated in a tissue culture laboratory, since these will harbour many contaminants which, in an open room, will inevitably find their way into the cultures.

Before culturing, it is essential that the hands are washed with a bactericidal soap such as is used by surgeons (e.g., Hibiscrub), whether or not sterile gloves are to be worn. Long hair should be tied back so that it does not fall over the cultures, and it is advisable to remove excess jewellery from the fingers and wrists such as large rings, loose bracelets and so on. The work-top area should be wiped down with 70% alcohol or an alcohol/bactericide disinfectant such as Chlorhexadine.

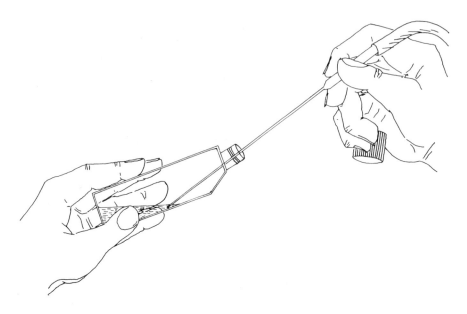

Figure 1. The correct method of holding screw caps while carrying out other procedures. The cap is unscrewed from the vessel by curling the fourth finger around it and gripping it to the side of the palm. This leaves the other fingers free to hold, for example, a pipette.

Mastering the various techniques involved in manipulating pipettes, syringes, culture vessels and other items of tissue culture requires care, concentration and practise under supervision. Numerous repetitions of operations such as removing a needle from a syringe without touching the plastic lip for support, or holding a number of objects between the fingers while carrying out a procedure may be necessary before this becomes automatic. It is important that bad culturing habits be noted and rectified early in training before they become difficult to break.

Figure 1 illustrates the correct way to remove a screw cap from a bottle or culture flask so that it can be held while other procedures are being carried out. This should be practised on non-sterile items until it becomes habit. It is important never to put down on the bench items that are to come into direct contact with the sample or culture until they are finished with.

When treating more than one culture at a time, it is essential that they are not cross-infected. This can be prevented by ensuring that items which will come into direct contact with the culture, medium or inside of culture vessels are used only once (for example, a fresh pipette must be used for each culture when withdrawing medium for renewal).

Culture vessels may accumulate dust while in the incubator, and removing them from a warm atmosphere causes a change in pressure which will force air, and consequently dust-borne contaminants, into the culture. This may be prevented by flaming the neck of the vessel with a small bunsen burner, which fixes the dust in place, thus preventing it from being introduced into the culture vessel. This may not be possible with designs of safety cabinet which do not allow for connection of a burner, but many do have this facility. Flaming is also useful if medium has leaked into the thread of the vessel

4

neck through overfilling or dropping. In this case the neck must be carefully wiped with a sterile swab, and flamed gently before discarding the old cap and replacing it with a fresh one. It is important that care be taken when flaming a plastic vessel because the neck may melt or warp so that the cap cannot be properly replaced.

Most hospitals or laboratories have strict rules regarding the disposal of pathological samples and potentially infected items which have been used for tissue culture, and it is essential that these are adhered to. Sharp objects including syringe needles, pipettes and scalpel blades are normally placed in a commercially available bin (e.g., Cin-Bin), while plastics and materials other than glassware are put into disposable bags which can be autoclaved or incinerated. Glassware is usually dealt with separately, and incinerated before disposal. Further information on the health and safety aspects of tissue culture may be found in references (1) and (2).

2.1.2 *Sterilisation*

All items that are likely to come into direct contact with the culture, or culture media, must be sterile, and this may be achieved in the following ways.

(i) *Irradiation.* Commercially available plastics for tissue culture are normally sterilised with gamma irradiation, but this is never performed in the routine cytogenetics laboratory. Alternatively, a short exposure to u.v. light may be used.

(ii) *Heat.* Either dry heat or moist heat may be applied. Dry heat is used mainly for apparatus such as glassware which is not damaged by high temperatures. An oven is normally used for dry heat sterilisation, and this must be pre-heated to the correct temperature, usually 160°C, and items exposed for 60 − 90 min. Moist heat is more suitable for materials which may be damaged by very high temperatures, and is usually applied in the form of steam under pressure using an autoclave. Most items will be adequately sterilised by autoclaving at 8 kg pressure and 115°C for 20 min, although solutions require only 15 min.

Paper bags and adhesive tape are commercially available which are resistant to heat and indicate, by a colour change, that the correct temperature for sterilisation has been reached. Items may therefore be sealed inside bags of this sort, and secured with the special tape, for sterilising, and may be subsequently opened inside the safety cabinet. It is advisable to insert a plug of cotton wool into the mouth of flasks, bottles and glass pipettes before sterilising to serve as a filter to the air which is inevitably sucked in as cooling takes place. Some items, such as metal instruments, may be sterilised by boiling.

Instruments which may be used to handle samples, such as forceps, may only require a rapid sterilisation of the tip. This can be achieved by dipping an appropriate length of the instrument into 70% ethanol and flaming until red-hot (it should be allowed to cool before handling the specimen!). Repeated flame-sterilising does, however, considerably shorten the life of the forceps.

(iii) *Filtration.* Liquid reagents which may be destroyed by heat can be sterilised by drawing through a filter of a pore size smaller than contaminating organisms, usually 0.22 or 0.45 μm. Commercially available filtration systems include those made by Millipore, and Flow Laboratories. When filtering biological material such as serum, pressure is preferable to suction as the latter may cause CO_2 to be produced which, in turn, alters the pH.

Most tissue culture reagents and apparatus can be obtained ready-sterilised from commercial suppliers. This minimises the time spent applying the above procedures, but may substantially increase the cost of such items. Consequently, many laboratories prefer to economise by doing their own sterilising.

2.1.3 *Antibiotics and Fungicides*

Assuming that all items for tissue culture have been sterilised, and that good aseptic technique has been applied (with the aid of a safety cabinet) the risk of infection in the culture should be minimal. Thus the use of antibiotics and fungicides in culture media acts as a 'safety net' should contaminants still be introduced, and these should never be relied upon as an alternative to good aseptic technique. Since antibiotics and, in particular, fungicides may have an inhibitory effect on the growth of human cells in tissue culture, their use should be kept to an absolute minimum, in terms of both the number of different varieties used and concentration. In addition, there is often a strong risk that contaminants may develop resistance to a particular antibiotic. This can be dangerous if the organism in question is a pathogen and could lead to infection in laboratory personnel which is difficult to treat.

In order to use antibiotics sensibly and effectively it is necessary to examine the organisms which may infect a culture.

(i) *Bacteria.* Common bacterial contaminants include Streptococci (e.g., α and β haemolytic streptococci), Staphylococci (e.g., *S. aureus*), *Pseudomonas* sp., and *Escherichia coli*. These are found in the normal skin flora, although they may occur in greater numbers during mild infections of the throat or nasal passages. Penicillin and streptomycin are the most commonly used antibiotics for tissue culture. Between them they are effective against a wide variety of bacteria and remain stable in tissue culture media at 37°C for up to 3 days, although the activity of penicillin may be depressed by up to 30% in the presence of serum. Kanamycin and gentamicin are also suitable since both have a broad antimicrobial spectrum. In addition, these two antibiotics are effective against some mycoplasmas, and remain stable in tissue culture media at 37°C for up to 5 days. Polymyxin B is effective against a range of Gram-negative bacteria and may be used if infection with a *Pseudomonas* sp. is suspected. The above antibiotics are all commercially available, prepared specifically for use in tissue culture, and are normally used at the the following final concentrations:

(a)   penicillin: 100 units/ml;
(b)   streptomycin: 100 μg/ml;
(c)   kanamycin: 100 μg/ml;
(d)   gentamicin: 50 μg/ml;
(e)   polymyxin 'B' sulphate: 100 units/ml.

(ii) *Yeasts and fungi.* With the exception of certain yeasts which are normal skin contaminants, most fungal infections are introduced from the air or from water. Common fungal contaminants include *Candida* sp. (e.g., *C. albicans*, yeasts found in body flora) and *Aspergillus* sp. (e.g., *A. fumigatus*, spores present in the air). The two fungicides commonly used for tissue culture are nystatin (Mycostatin, Squibb) and amphotericin B (Fungizone, Squibb). Of these, nystatin is less detrimental to cultured human cells than amphotericin B, and should be used in preference where possible. Both fungicides remain stable in tissue culture media at 37°C for up to 3 days and are used at the following final concentrations:

(a)   nystatin: 50 μg/ml;
(b)   amphotericin B: 2.5 μg/ml.

(iii) *Mycoplasma*. Both gentamicin and kanamycin have some anti-mycoplasmal activity.

(iv) *Viruses*. Viruses live intracellularly without killing their host cells. Many of these organisms arise in primary cultures and are usually present as chronic infection, at source, in the tissue in question. In general, if cells become infected with virus, very little can be done.

2.1.4 *Outbreaks of Contamination*

All laboratories have suffered outbreaks of contamination from time to time. It is very important to attempt to trace the source of the outbreak, and rectify it if possible. It is essential that the contaminating organisms are identified, as this will show whether the source is environmental or human. This will normally involve the services of the bacteriology (and mycology, if there is one) department who will usually be able to identify the organism from a sample of the infected culture within a few days. Common sources of contamination are given below.

(i) *Specimens*. Particularly post-mortem material such as abortus or skin samples. These may be put into culture before infection is evident, which can lead to cross-infection of other cultures in the incubator. It is advisable to keep cultures from 'clean' specimens such as amniotic fluids in a separate incubator from 'risk' specimens such as products of conception.

(ii) *The operator*. Especially if he has a cold or a mild throat infection. It is advisable to avoid culturing while suffering from any mild ailment which may increase the risk of infecting cultures. Infection with microbes present in body flora will almost certainly have originated from the operator, and may have resulted from bad aseptic technique.

(iii) *Faulty safety cabinet*. These should be serviced regularly, and the filters changed. A hole in a filter may result in contaminated air being forcibly blown into cultures.

(iv) *Contaminated media or apparatus*. This sometimes occurs in faulty batches of commercially supplied items, or may result from insufficient sterilisation (for instance, trypsin may sometimes be infected with yeast whereas Gram-negative bacteria and mycoplasmas may occur in sera). Contaminants may even occur in antibiotic solutions.

(v) *Concentrated contamination source in the laboratory*. Such as air conditioning, dirty laboratory conditions, water bath or sink. This has already been discussed in Section 2.1.1.

(vi) *Incubator*. Humidified gas incubators present a high risk of contamination, and this is discussed in Section 2.5.3.

Any cultures, media or reagents which are found to be infected should be discarded as soon as this has become evident. Bacterial infections usually manifest rapidly, giving a cloudy appearance to the medium which, if it contains a pH indicator such as phenol red, will have turned yellow reflecting an acid pH. Yeasts may be present for several days before becoming visible, especially if a fungicide has been used. These also give the culture a cloudy, yellow appearance but a microscopic examination with a $\times 100$ invertoscope will reveal larger microbial cells than would be seen with bacterial contamination. Fungal contamination is normally first noticed when a large furry colony has appeared in the culture. Microscopic examination of other cultures will reveal the presence of any filamentous mycelia which may be present. Special kits are commercially available for detecting mycoplasma infection, but other than this there is no way to be sure. Symptoms of viral and mycoplasmal infection are, in general, a tendency for cells to look 'unhealthy' and for unusual cell types to appear in the culture which do not produce actively-growing colonies.

The following data should be recorded for each batch of reagents or sterile items to be used.

(i)   Manufacturer's batch number and date of receipt of all deliveries.
(ii)   For reconstitution of media and reagents, the name of the operator, date and batch numbers of all components used.
(iii)   For cleaning and sterilisation of equipment and utensils, the name of the operator, date, and brief details of the procedure used.
(iv)   Date of opening if the items in question are intended for further use.

The above records will be invaluable if it becomes necessary to trace the source of a contamination outbreak, but despite stringent steps to isolate the exact source of the contamination outbreak, it may never be discovered. If, however, the above precautions are taken, and all infected cultures are disposed of, the outbreak should eventually disappear. There is, of course, always the risk of recurrence, and outbreaks should always be recorded in case a pattern should emerge. Outbreaks may be associated with activities such as decorating which may be taking place nearby, or may have a seasonal influence.

2.2 Culture Vessels and Substrate

The substrate upon which cells are to be grown is of great importance, since even the slightest toxicity will have a damaging effect. It has been shown that cells grow best on a negatively-charged surface and many commercially available tissue culture vessels have been treated in some way to facilitate this.

Tissue culture vessels may be made of glass or plastic. Only borosilicate or soda glass should be used for tissue culture, since these are less toxic to cells than other types of glass which may contain lead. Borosilicate glass vessels are re-useable if properly cleaned and sterilised, and are cheaper than commercially available disposable plastic vessels. Polystyrene plastic vessels, on the other hand, are labour-saving and may provide a greater uniformity of surface chemistry. Most cytogenetics laboratories use some or all of the following four types of culture vessel.

2.2.1 Flask

Figure 2 illustrates the 25 cm^2 plastic flask with screw cap, which is the size normally used for primary cultures and early subcultures. Larger flasks (80 and 175 cm^2) may be used for culturing large numbers of cells (e.g., for freezing). Caps may be loosened for use in a CO_2 atmosphere.

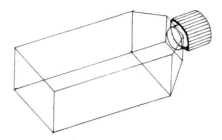

Figure 2. The 25 cm² tissue culture flask.

Figure 3. The disposable plastic Leighton tube with removable coverslip.

2.2.2 *Leighton Tube*

Figure 3 illustrates the Leighton tube (120 × 16 mm). These are often used as an alternative to flasks when a smaller surface area is an advantage, particularly in cases where several parallel cultures are required from a limited amount of specimen (e.g., amniotic fluid). Another advantage of the Leighton tube is that a coverslip can be placed along the flat surface if *in situ* harvest is required. Caps may be loosened for use in a CO_2 atmosphere. Particular care should be taken not to over-fill these vessels since they are especially prone to leakage.

2.2.3 *Petri Dishes and Multiwell Cluster Plates*

Figure 4a and *b* illustrates the 4-well multidish (well diameter 16 mm) and a 6-well multidish that may be used for tissue culture. Petri dishes and multiwell cluster plates (normally 35 × 10 mm) may be used with coverslips (22 mm²) for *in situ* preparations. These are supplied with vented lids for use in a CO_2 atmosphere.

Glass coverslips for use in Leighton tubes or Petri dish cultures must be cleaned and sterilised by the procedure described in *Table 1*. Alternatively, ready-sterilised polystyrene or Thermanox coverslips are commercially available (Flow laboratories). Glass bottles and pipettes should be cleaned and sterilised using the procedures given in *Table 2*.

2.2.4 *Flaskette and Slide Chambers*

Figure 5 illustrates the slide chamber (2.16 × 4.57 cm). Chamber vessels provide a way of enabling *in situ* harvest to be performed without using 'open' system tissue culture, such as is necessary with dishes containing coverslips. The culture is initiated within a small area on a sterile, and surface-treated microscope slide with a chamber sealed to it. Medium can be renewed through the neck of the 'flaskette' or by removing the lid of the slide chamber. When the culture is ready for harvesting, the chamber can be removed, and the slide treated in the same way as a coverslip culture.

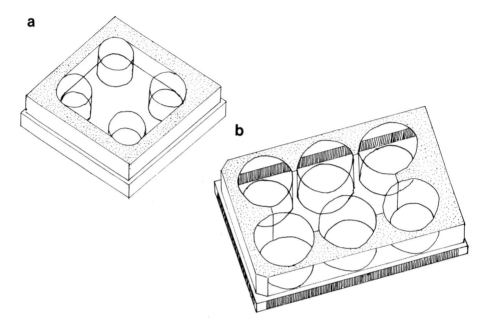

Figure 4. (a) The 4-well multidish. **(b)** The 6-well multidish.

Table 1. Preparation of Glass Coverslips for Tissue Culture.

1.	Prepare an acid/alcohol solution by adding 6 ml of concentrated HCl to 200 ml of 70% ethanol (treat this solution with extreme caution as it is very corrosive).
2.	Immerse the coverslips in 200 ml of acid/alcohol solution for at least 1 h but not more than 24 h.
3.	Drain off the acid/alcohol solution and rinse the coverslips thoroughly in distilled, de-ionised water.
4.	Wash the coverslips in distilled, de-ionised water a further nine times.
5.	Drain each coverslip individually on non-fibrous absorbent paper, such as Post-Lip, and lay flat to dry. Do not allow a water droplet to form on the coverslip or it will dry leaving a ring-shaped mark which will interfere with cell growth. If this should happen, carefully blot the coverslip without breaking it.
6.	Place dry coverslips between two filter papers (Whatman Grade 1) in a glass Petri dish according to the number required at a time.
7.	Wrap each Petri dish in an autoclave bag, seal and sterilise either in an oven or in an autoclave.

2.3 Media and Additives

The basis of all culture media is the balanced salt solution. This contains a combination of salts and glucose which control pH and osmotic pressure while providing an energy source. Cells can survive for a brief time in such salt solutions, but if they are to be cultured for longer periods it is necessary to include in the medium amino acids, vitamins, oxygen and serum proteins.

2.3.1 *Balanced Salt Solutions*

Balanced salt solutions are of two types:

(i)   those which equilibrate with air [e.g., Hanks' balanced salt solution (HBSS)]

Table 2. Cleaning and Sterilisation Procedures for Tissue Culture Glassware.

Bottles

1. Remove residual cell material from the bottle by rinsing in 10% sodium hypochlorite solution.
2. Rinse the bottle in tap water.
3. Soak overnight in a detergent recommended for tissue culture glassware such as Pyroneg, Decon or 7X.
4. Rinse repeatedly in hot, then cold, tap water.
5. Rinse at least three times in fresh distilled water (or by machine if available).
6. Dry, assemble with caps and so on, and dry-sterilise.

Pipettes

1. Soak the pipettes overnight in 10% sodium hypochlorite solution for decontamination.
2. Rinse well in tap water.
3. Soak in detergent as above for $3-4$ h.
4. Rinse as above.
5. Plug if necessary, and dry-sterilise.

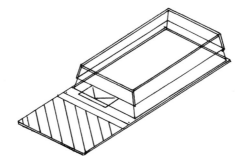

Figure 5. The slide chamber.

(ii)   those which equilibrate with a gaseous phase containing a high CO_2 tension, normally 5% [e.g., Earle's balanced salt solution (EBSS)].

In addition to forming the basis of media, balanced salt solutions are used as a base for other reagents such as trypsin, versene and colcemid. Dulbecco's phosphate-buffered saline (PBS) is often used for this, though whichever balanced salt solution is selected for reconstituting these reagents must be free from calcium and magnesium ions if the reagent is to be used for cell dissociation treatments.

Media suitable for the long-term survival of cells in culture must consist of a base medium, a buffering system, serum and a number of other additives.

2.3.2 *Base Media*

There are a number of base media in common use in human cytogenetics laboratories, most of which were originally developed for specific purposes, but which have now found wider application.

(i) *Medium 199 (TC 199).* Originally developed to study the nutrient requirements for cell survival and multiplication, this medium is now widely used for culturing most human tissues for chromosome analysis. Inclusion of EBSS assists longer survival of cells, although it may be alternatively used with HBSS depending on the buffering system employed. The depleted folate levels in this medium have led to its use in the study

11

of fragile sites on chromosomes, and a folate-free formulation is commercially available which, when used in conjunction with a low serum level ($2 - 4\%$), is suitable for cases intended for fragile X analysis.

(ii) *Minimal essential medium (Eagle's) (MEM).* A simpler medium than 199, MEM was formulated some years later to study further the nutritional requirements of mammalian cells. It has now become a standard basic growth medium, and contains either HBSS or EBSS according to the buffering system required. Various mixtures of non-essential amino acids and vitamins are available as supplements. Two major modifications to this medium exist, the Alpha modified MEM, which contains both essential and non-essential amino acids, and Dulbecco's modified MEM, which is a much richer medium due to the inclusion of amino acids and vitamins in concentrations four times that of basic MEM.

(iii) *Ham's nutrient mixtures F10 and F12.* The Ham's balanced nutrient mixtures were designed for the study of clonal growth requirements of diploid Chinese hamster and human cell lines, but are particularly useful for the growth of fastidious cell lines. Ham's F10 is particularly favoured for amniotic fluid culture, since poor samples which may not grow well in TC 199 or MEM often fare better in this medium. Ham's F12 was developed as a modification of F10 for cloning and serial propagation of Chinese hamster cell lines in the absence of serum, but is normally used with serum for human tissue cultures.

(iv) *RPMI 1640.* This medium was originally designed for the cultivation of leukaemic cells and the long-term culture of peripheral blood lymphocytes. It has now become a general purpose medium and may be used for amniotic fluid cell culture. There are two modifications of this medium, Dutch and Searle, which are claimed to be more suited to lymphocyte stimulation experiments and which have different buffering systems. The low thymidine content makes RPMI 1640 a suitable medium for methotrexate, fluorodeoxyuridine (FUdR) and thymidine-block cell synchronisation cultures.

(v) *Leibovitz L-15.* This medium was designed for the growth of human fibroblasts in the presence of serum but without CO_2.

(vi) *Chang medium.* This highly complex and rather unstable medium has been recently formulated for the rapid growth of amniotic fluid cell cultures. Furthermore, it has found application in the newly-developed chorionic villus culture methods. Manufactured exclusively by Hana Biologics, it is considerably more expensive than the above media, mostly due to inclusion of growth factors and hormones, and has a very short shelf life (3 months). It does, however, have a dramatic effect on fetal cell cultures and can lead to as much as a 50% reduction in the time taken to harvest. In addition, it can be used to stimulate cultures which are growing badly in conventional media, and will induce growth from aborted products of conception which may not otherwise be obtained. There are two versions of the formulation, one optimised for 'closed' culture systems (Chang C), the other suitable for both 'closed' and 'open' systems (Chang A). Since a small amount of serum is already included in the ingredients, this does not have to be added, but L-glutamine and antibiotics must be added as with other media. Chang A and Chang C supplements, which contain the unstable factors of the medium, are supplied separately from the base medium (Chang B). Supplements A and C are supplied either lyophilised (for reconstitution with water) or frozen, while supplement B is either in liquid or powdered form. The medium is reconstituted A + B or C + B immediately prior to use, and will remain stable for 5 days only. It is, therefore, normally convenient to aliquot Chang A or C solutions into convenient amounts which can be stored at −20°C. Chang B may be supplemented with L-glutamine and antibiotics, and stored in a refrigerator for several weeks if necessary, and an appropriate amount added to frozen A or C aliquots immediately prior to use. Complete Chang medium cannot be stored frozen, neither should A and C solutions be refrozen once thawed.

It is beyond the scope of this chapter to include detailed recipes of the above media, since few laboratories constitute these themselves, but most catalogues produced by commercial suppliers include the exact formulations of their media (see Appendix). Commercially obtainable media are subjected to stringent quality control which includes the testing of sterility, functional properties, biochemical and physical properties and endotoxin levels.

Media and balanced salt solutions may be supplied either in liquid or powdered form. Liquid media may be single-strength and ready for use, or be supplied in 10× concentrate for a more economical bulk purchase which requires diluting. Some powdered media suitable for autoclaving may be obtained. Water for reconstituting powdered media and diluting concentrated liquid media should, ideally, be purified by de-ionisation, reverse osmosis and ultra-fine filtration, and monitored for acceptable levels of purity in terms of chemical elements and endotoxins. Reconstitution of concentrated or powdered media must be carried out according to the supplier's instructions after which they can then be filter-sterilised (or autoclaved in the case of specially formulated powders).

L-Glutamine is normally added to media shortly before use since it is unstable at room temperature or above. It is stored at −20°C. Some media may already be supplemented with L-glutamine, but this should still be added fresh to ensure maximum effectiveness.

2.3.3 *Buffering Systems*

The growth of animal cells in a nutritionally complete tissue culture medium is optimum when the medium is buffered at pH 7.2 − 7.4. In addition to the buffering effects of the balanced salt solutions already discussed, two other buffering systems are in common use.

(i) *Sodium bicarbonate.* Although sodium bicarbonate is a nutritional requirement of most human cell lines, its use as a buffering system is limited by two factors. Firstly, its pK_a is 6.3 at 37°C resulting in suboptimal buffering throughout the physiological range, and secondly it causes CO_2 to be released into the atmosphere during metabolism thus increasing the alkalinity of the medium. The production of CO_2 may be controlled, however, by artificially supplying CO_2 to the atmosphere to prevent the gas leaving the liquid. Sufficient buffering may be provided in some cell culture regimes by using an Earle's salts-based medium together with 10 mM sodium bicarbonate and, in this case, there is no need for CO_2 gassing.

(ii) *Hepes buffer.* Hepes buffer does not require a CO_2-enriched atmosphere, and the pH reading of medium in this buffer varies inversely with temperature. Therefore,

13

Hepes-buffered media at 15°C with a pH of 7.5−7.6 will have a pH of 7.2−7.35 at 37°C, and this should be taken into account when media are prepared. The pH of freshly prepared media may be adjusted with 1 M NaOH or 1 M HCl as appropriate. Although this buffer does not actually maintain a constant pH, it tends to stabilise and resist rapid pH changes in the medium. The presence of sodium bicarbonate greatly influences these pH changes and if this is to be used together with Hepes buffer then the concentration of each should not exceed 10 mM otherwise the tonicity of the medium may be raised to an unacceptable level. 20 mM Hepes buffer is otherwise used and this is suitable for use with either Hanks' or Earle's salts.

2.3.4 *Sera*

The precise role of serum in the complete cell culture medium is still not fully understood but it is evident that, without it, or some form of substitute, cells do not survive in culture for long. Cytogenetic laboratories use either bovine or human sera for human tissue cultures, both of which may be obtained commercially. Three types of bovine sera may be used, fetal calf serum (FCS), newborn calf serum (NCS) (taken from animals 10 days old or less) and calf serum (CS) (taken from animals 10 weeks old or less). By far the most effective, but also the most expensive, is FCS. Although human AB serum is commercially available, it is cheaper to prepare it from pooled serum obtained from blood banks if at all possible. The preparation of serum is a relatively simple procedure, which consists of allowing whole blood or plasma to coagulate, and subsequently removing the exuded serum. The serum then has to be filtered and sterilised. Sera should be stored frozen at −20°C.

Serum must be added to all the above-mentioned media, with the exception of Chang medium, at concentrations varying from 10 to 30% depending on the type of culture. A compromise of 20% is probably most beneficial, because an excess of serum can prove detrimental to the cells, and a shortage of serum will not allow maximum growth to be attained. It is the most variable component of medium, and because it is a biological fluid it may be infected with microorganisms, particularly mycoplasma. For this reason commercial suppliers apply stringent quality control measures which analyse levels of various metabolites, proteins and endotoxins as well as sterility checks, particularly for the detection of mycoplasmas. Most suppliers allow laboratories to batch-test sera before purchasing a bulk supply, and full advantage should always be taken of this facility. Batch-testing in the laboratory is normally carried out by comparison of cell growth in medium containing the test serum and one containing a serum that is known to perform well in culture. Alternatively this may be done with a radiomicroassay method (3).

Gibco Ltd. now produce a serum, 309, which has very low endotoxin levels, and which is subjected to more tests than are normally performed on other sera. This enables a 'batch-matching' system which is of considerable advantage if a previous batch has performed particularly well in routine use.

(i) *Serum substitute − Ultroser G.* Ultroser G has recently been developed as a serum substitute by IBF reactifs, and is marketed by LKB. Included in its formulation are those factors identified in serum as being essential for cell growth, such as growth factors, adhesion factors, mineral trace elements, hormones, binding proteins and vitamins.

A trypsin-inhibiting factor is also included in the formulation. Despite claims that batch variation is not a problem associated with Ultroser G, some laboratories have found that this does occur, and some artefactual chromosome changes have been noted as a result of its use. Many laboratories have, however, found Ultroser G to be suitable for use in culture media such as Ham's F10 if used with a very low concentration of FCS (5%).

2.3.5 *Medium and Sterility*

Once the various components have been added to the medium, or it has been reconstituted from concentrated or powdered forms, it is essential to check the sterility. This may be done by removing a small aliquot from each bottle of medium to be tested, and sending samples to the bacteriology laboratory for analysis. Alternatively, the aliquots may be incubated at 37°C overnight or for several days before the medium is required for use. Contaminated medium will become cloudy in appearance, and turn yellow in colour, if phenol red is used as a colour indicator, denoting an acid pH. A log book should be kept for recording details of all media and reagents used, as discussed in Section 2.1.4.

2.4. **Temperature, pH, Osmotic Pressure and Gas Tension**

2.4.1 *Temperature*

Since 37°C is the optimum temperature required by human tissues and cells, cultures must be placed in an incubator or hot room that can maintain this temperature without fluctuations of more than ± 0.25°C. Increasing the temperature affects the cellular multiplication rate; the higher the temperature, the greater the rate of growth, until a certain temperature limit is reached. When the upper limit is exceeded, then any further increase in temperature is inhibitory to the cell growth. Lower temperature has the effect of depressing cell metabolism, but this does not appear to affect the cells in a deleterious way provided that all other factors are still maintained. It is important, therefore, to monitor carefully the temperature within the incubator, or hot room, using a thermometer in addition to the gauge on the indicator panel if there is one.

2.4.2 *pH*

The question of pH has already been discussed in relation to buffering systems which can be applied to culture media. pH is also influenced by the gaseous environment surrounding the culture, and this in turn must be considered together with other factors such as temperature, osmotic pressure and gas tension.

2.4.3 *Gas Tension*

The role of CO_2 in tissue culture is discussed in Sections 2.3.3 and 2.5. Oxygen is also required by cells. A pressure of $15-75$ mmHg is usually adequate, higher concentrations being inhibitory to cell growth. Thus 'gassed' cultures are normally exposed to a 95% air/5% CO_2 mixture.

2.4.4 *Osmotic Pressure*

The optimum range of osmotic pressure for growth is narrow, and varies with cell type.

Fetal fibroblasts, for example, have a range of $250-325$ mosmol/kg. Thus, when using a large concentration of Hepes buffer, an adjustment of NaCl concentration is necessary.

2.5 **Incubators**

The choice of incubator or hot room will depend on the cell culture system that is to be used, and these requirements can be considered in two categories, 'closed' or 'open' systems.

2.5.1 *Closed Systems*

A closed culture system is one which does not require circulated gas mixtures or humidification, since culture vessels are air-tight. This includes flask and Leighton tube cultures with screw caps tightly sealed, and Flaskette chambers. Bicarbonate-buffered closed cultures require a high CO_2 tension, and this is supplied by 'gassing' the cultures from a small cylinder of compressed CO_2. Alternatively, cultures may be placed in a 'sandwich box' which is then gassed and incubated. Flasks and Leighton tubes are incubated in such boxes with their caps loosened, as in open-system culture, or vented Petri dishes or multiwells may be used (strictly speaking, this then becomes a 'semi-closed' system). A disadvantage of the gassing system is that over-acidification may occur due to the accumulation of CO_2 produced in the metabolic process. Gaseous contaminants may occasionally arise, but they can be combated by in-line filtration. Hepes-buffered closed cultures do not require gassing.

Closed systems are, on the whole, preferred by many cytogenetics laboratories since they do not require elaborate gas regulation and humidification systems with all the problems that these may cause. This enables the use of cheaper and simpler incubators than the open system, since all that is required is a good temperature-control system. Alternatively, closed cultures can be grown in a specially designed hot room.

The closed system is particularly suited to cultures intended for 'suspension harvest' by enzymatic and physical removal of cells from culture vessels. *In situ* harvest can, however, be performed on coverslip cultures grown in closed Leighton tubes or on slide cultures from Flaskette chambers.

2.5.2 *Open Systems*

An open culture system is one which allows gaseous interchange between the air inside the culture vessel and the surrounding atmosphere. Multiwell and Petri dishes with vented lids must be used in an open system with humidification to prevent evaporation of medium, since this will alter the osmolarity. Open flask and Leighton tube cultures may also require humidification if small amounts of medium are to be used, and caps must be loosened by one turn of the thread to allow air circulation into the cultures. Incubators designed for use with bicarbonate-buffered open cultures must be able to regulate a supply of CO_2 from cylinders of compressed gas to a level of 5% and if humidification is required this should be maintained at 97%. Hepes-buffered open cultures do not require 5% CO_2.

The main advantage of the open system for human tissue culture is that *in situ* harvest methods can be applied. In addition, it is possible to use smaller culture vessels, such as microwells, which is of particular advantage to chorionic villus culture. There are, however, several serious disadvantages inherent in the use of open culture systems with CO_2 and humidification. Firstly, this combination is conducive to the growth of microbial contaminants, paticularly fungi, which may be harboured in the tubes and reservoirs of the humidification system, or on the damp surfaces of the incubator. Secondly, open-system cultures are more easily infected than closed cultures, and there is a greater risk of large-scale contamination outbreaks. Additionally, a greater sophistication of incubator equipment is required which is not only more expensive, but introduces a dependence on more factors. For example, although most incubators can be fitted with an automatic changeover unit to enable two gas cylinders to be used, one in service and one in reserve, in practice these may malfunction with disastrous effects if the pH of the culture medium is increased at a critical stage for an extended period (e.g., amniotic fluid cultures less than 5 days old may be severely affected if CO_2 cylinders expire during a weekend). Although in the light of these disadvantages it seems unwise to use an open system with 5% CO_2 and 97% humidity, some cytogeneticists believe that better cell growth is obtained from 'weak' or difficult cultures than is possible with closed, Hepes-buffered systems.

2.5.3 *Humidification*

Closed culture systems require a simple, temperature-controlled incubator. There are several designs of CO_2 incubator with humidification control suitable for open culture systems, although some of these may be more subject to problems with microbial con-tamination than others. The simplest way of humidifying an incubator is to place a trough of water on the lowest shelf, although some designs incorporate a removable tray. If such trays are autoclaved regularly, and filled only with sterile water (tap water or distilled water contains large quantities of contaminants which, once introduced into an incubator, are almost impossible to get rid of) then this humidification system is effective and relatively harmless. Sodium benzoate may be added to this water to reduce the risk of yeast and fungal infection since it is an innocuous chemical. Fungicides, antibiotics and volatile bactericides (including Panabath) must never be added to water intended for use in an incubator humidification system.

A particularly dangerous style of humidified incubator in terms of contamination risk is that which incorporates a large vat of water, which is placed outside and above the incubator, connected by means of a tube to a reservoir with a ballcock, attached to the outer wall. This water passes into the incubator where it is heated, the steam pro-viding the required humidification. This system is particularly prone to fungal infec-tion which, once it has been introduced, is impossible to eliminate. Fumigation pro-cedures involving volatile substances cannot be applied to incubators. Such volatile substances may cause damage to the incubator and may also be detrimental to open cultures.

In our experience, the safest and perhaps the most efficient design of humidification system is that of Assab. Sterile water is placed in two vats attached to the inside of the inner incubator door. The air/CO_2 mixture is passed through this water by means of two tubes leading from a pump which is automatically regulated. The components of the humidification system are also removable and autoclavable so that infection, if it does occur, can be eliminated.

Water for humidification should be at the same temperature as the inside of the incubator. Opening the incubator causes an in-rush of dry air, and this cools the water by evaporation, and the water may remain cool for a prolonged period inadequate for the prevention of further evaporation. Regular inspection of the incubator should guard against this, by checking the insulation round the door, and if inspection is made regularly, then this should prevent any cold-spots appearing. Damaged insulation serves as a site of condensation and also prevents the partial pressure of water vapour reaching the desired humidity.

More detailed information on the above aspects of tissue culture may be found in references (4) and (5) which are good basic textbooks on general tissue culture methods. We will now consider the culture of specimens specifically intended for chromosome analysis.

3. AMNIOTIC FLUID CULTURE

Amniotic fluid is obtained by amniocentesis performed at or after 14 weeks into pregnancy. The optimum amount required for cell culture is 20 ml, preferably divided into two 10-ml aliquots so that parallel cultures may be set up in different media, which serves as an 'insurance' against faulty or infected batches of medium. Samples received in one container only should be divided into two aliquots and then centrifuged. Smaller samples are, however, frequently received but the chances of obtaining a successful culture from amounts of 5 ml or less depends on the number of viable cells present. This, in turn, is dependent on factors such as gestational age and state of the specimen (e.g., bloodstained, brown etc.). The interval between sampling and receipt of the specimen by the laboratory should also be noted. Amniotic fluid can survive for 24 – 48 h if refrigerated, but should *always* be set up fresh where possible. Thus it can be transported some distance, but should be received within 24 h of sampling to ensure a maximum chance of success.

The specimen should be received in a sterile vessel with a conical base so that it may be centrifuged directly without the need for transfer into another vessel (thus minimising contamination risk and loss of cells). Note should always be made of the amount of specimen received, appearance (e.g., heavily bloodstained, 'pink' and therefore slightly bloodstained, brown) and the size of the pellet after centrifugation. A recommended medium for amniotic fluid culture is detailed in *Table 3*. All other stock solutions used are detailed in *Table 4*.

3.1 **Preparation of Amniotic Fluid Cell Cultures**

3.1.1 *Setting up the Culture*

(i)   Centrifuge the specimen at 1000 r.p.m. for 10 min.

(ii)   Remove the supernatant from each aliquot, with a syringe or pipette, leaving about 0.5 ml of the fluid above the undisturbed pellet.

(iii)   Transfer the supernatant into a universal container, for α-fetoprotein (AFP) and acetylcholinesterase (AchE) estimations.

(iv)   Resuspend each pellet in an appropriate amount of medium for the type and number of culture vessels to be used and seed the cell suspension as shown in *Table 5*.

Table 3. Recommended Complete Medium for Amniotic Fluid and Solid Tissue Cultures.

Material	Strength	Volume (ml)
Ham's F10	1×	100.0
Fetal calf serum	−	20.0
L-Glutamine	200 mM (100×)	1.0
Penicillin +	10 000 IU/ml +	1.0
Streptomycin	10 000 µg/ml	
Nystatin	10 000 IU/ml	0.5
Sodium bicarbonate	7.5%	1.0
Hepes buffer (if appropriate)	1 M	1.0

Table 4. Tissue Culture Reagents − Stock Solutions.

Reagent	Stock solution
Trypsin	0.25% (1:250) in normal saline
Versene	1:5000 in isotonically buffered saline
Colcemid	10 µg/ml either prepared in HBSS or lyophilised in PBS for reconstitution with water

Table 5. Seeding Volumes per 10-ml Amniotic Fluid Cell Suspension[a].

Vessel	Vessel size	No. of vessels	Vol. of medium to add to cell suspension (ml)	Seeding vol. per vessel (ml)
Flask	25 cm²	2	10.0	5.0
Leighton tube (coverslip)	16 × 120 mm 9 × 35 mm	3	4.5	1.5
Petri dish or multiwell (coverslip)	35 × 10 mm 22 mm²	3	6.0	2.0
Multiwell (coverslip)	16 × 11 mm 13 mm diameter	4	3.0	0.75
Flaskette (1 chamber)	2.16 × 4.57 cm	2	8.0	4.0

[a]Each well of a multiwell cluster plate is considered as a separate vessel in this table.

(v)   Incubate the seeded cultures at 37°C, with CO_2 and humidification if appropriate to the culture system used (see Section 2.5).

3.1.2 *Culture Maintenance*

Cultures should be left undisturbed for at least 7 days, after which they should be examined with a ×100 invertoscope for cell growth. If cells have already begun to settle and colonies are starting to form, about half of the medium should be removed and replaced with fresh medium. If, however, only a few isolated cells have settled, this half-medium change should be left until more growth has taken place. Further, complete, medium changes should be peformed twice a week until the culture is ready for harvesting. Alternatively some laboratories prefer to sub-culture before harvesting. This is achieved by trypsinising cells into suspension and aliquoting appropriate amounts into the required number of vessels. The resulting sub-cultures may be sub-cultured further and each successive culture is referred to with a 'passage level' number. Amniotic fluid cell cultures are, on the whole, unable to tolerate more than a few passages. Fibroblasts, on the other hand, will tolerate many more.

Cultures seeded in Chang medium should be examined after 5 days, and a complete medium change performed because this medium is unstable after that time.

3.1.3 *Harvesting*

(i) *Non-coverslip cultures.* These are ready for harvesting when four or five confluent colonies are present (typically 8 – 14 days after seeding). The most popular harvesting method in cytogenetic laboratories is the 'trypsin-suspension' method, although flasks can also be harvested by the method described for chorionic villus cultures (Section 4.3.1).

(i)   Inoculate the culture with sterile colcemid solution (*Table 4*) to a final concentration of 0.04 μg/ml.

(ii)   Incubate the culture for 4 h at 37°C.

(iii)   Remove the culture medium and replace with 1 ml of versene solution (*Table 4*).

(iv)   Wash the cells thoroughly with the versene to mix with the residual medium, and remove by suction. This ensures that no serum is left in the culture vessel that could inhibit the trypsin action in the next stage.

(v)   Inoculate the culture with trypsin solution (*Table 4*), 1 ml per flask, 0.5 ml per Leighton tube and 1 ml per Petri dish.

(vi)   Ensuring that the cap is tightly screwed, incline the flask or Leighton tube slightly downwards and repeatedly tap the side firmly to dislodge cells. Check the culture with an invertoscope at this stage. Cells should be 'rounding up' (i.e., the cells which normally have a flattened, elongated appearance will be seen taking on a rounded three-dimensional appearance). Apply several more firm taps to the flask and check again. Cells should now been seen detaching from the surface of the flask and floating in the medium. Cells are dislodged from Petri dish cultures by repeatedly syringing the trypsin solution up and down. Some laboratories combine the trypsinisation treatment with the physical removal of cells by means of a 'rubber policeman' (rubber-capped glass tube) which is used to scrape cells off the surface of the flask. This may, however, lead to clumping and damage of the cells and we do not recommend this approach.

(vii)   Arrest the trypsin action by adding 5 ml of culture medium (the serum in the medium contains α_1-anti-trypsin.

(viii)   Transfer the cell suspension, aseptically, into a conical-based centrifuge tube (siliconise glass tubes by the procedure given in *Table 6*) and centrifuge at 1000 r.p.m. for 10 min. The flask, tube or Petri dish can then be replenished with medium and re-incubated, or a fresh vessel can be inoculated with some cell suspension prior to centrifugation, medium added as above and the culture subsequently maintained. The medium should be replenished on the following day to remove all traces of trypsin. Thus the culture can be perpetuated.

(ix)   Remove the supernatant and discard. Resuspend the cell pellet in 10 ml of

Table 6. Procedure for Siliconising Glass Tissue Culture Utensils.

1.	Reagent: 'Repelcote' (a 2% solution of dimethyldichlorosilane in 1,1,1-trichloroethane).
2.	Carefully apply Repelcote to the surfaces to be siliconised so that it is evenly spread. Repelcote may be poured into tubes, or pipettes can be laid horizontally in a dish so that the solution soaks into the bore. Excess solution can be put back into the original bottle and re-used. This must be done in a safety cabinet since the fumes are dangerous if inhaled.
3.	Leave to dry for at least 2 h. Invert vessels such as centrifuge tubes over Post-Lip paper to collect excess solution.
4.	Apply the washing procedure described in *Table 2* to remove traces of HCl.

0.075 M KCl mixing thoroughly either with a pipette or a 'Whirlimix'. Leave to stand for 15 min.

(x) Centrifuge the tube at 1000 r.p.m. for 10 min.

(xi) Remove the supernatant, leaving about 0.5 ml of the KCl above the pellet and slowly add 10 ml of freshly made fixative (3:1 methanol/glacial acetic acid), mixing throughout to avoid the formation of lumps.

(xii) Repeat the fixative washes twice more. Remove the supernatant after the last wash and add about 0.5 ml of fixative to the cell pellet, depending on size.

(xiii) Prepare the slides. There are several ways of doing this, but probably the easiest is the 'drop' method. Slides should be clean and grease-free to ensure good spreading, and the best way to achieve this is by soaking first in absolute methanol, and then in de-ionised water. Slides can be stored in water and used wet or dry depending on preference. A small amount of cell suspension is drawn into a Pasteur pipette and about three drops are expelled carefully in three different positions on the slide. Spreading is achieved by the movement of the periphery of the drops outwards until dry.

Variations of this spreading method include using slides which have been stored dry at −20°C, spreading onto slides placed on a block of dry-ice, flame-drying after dropping and dropping from different heights. Cell suspension may also be applied to slides by placing a large drop at the top of the slide and blowing it vigorously downwards to the end of the slide. The slide is then dried on a hot-plate at about 40°C.

The quality of spreading may be influenced by temperature; high temperatures may cause over-spreading and cell breakage, whereas low temperatures may inhibit spreading. This is due to the different rates of evaporation of the fixative. These effects may be counteracted by the application of the above variations (in the case of low ambient temperature), or by placing the cell suspension at −20°C for an hour or so before spreading (in the case of high ambient temperature).

(ii) *Coverslip cultures.* Depending on the size of the coverslip used, cultures are ready for harvesting when three or more small colonies have been established (typically 6 − 10 days after seeding). On the whole, colonies should be less confluent for coverslip harvests than for suspension harvests, since overcrowding of cells will obscure the metaphase spreads and inhibit mitosis. Most of the useable mitoses will occur around the periphery of the colony, and these are easily identified by their 'rounded up' appearance, similar to that of trypsinised cells. Hence, harvest is indicated if several actively dividing colonies are apparent.

(i) Transfer the coverslip to be harvested to a fresh dish or multiwell. Replenish the medium in the original dish to maintain cells which may have grown over the coverslip onto the bottom of the dish.

(ii)   Add medium to the fresh dish (using the same volumes as for seeding) and incubate for 24 h.

(iii)   On the following morning, inoculate the dish with colcemid solution to a final concentration of 0.04 μg/ml. Incubate for 4 h at 37°C.

(iv)   Remove the medium by suction and carefully replace with an equal volume of 0.075 M KCl. It is important that reagents are added gently since the rounded-up mitoses are easily detached from the coverslip.

(v)   After 30 min, carefully add an equal volume of freshly made 3:1 fixative. This is best achieved by running fixative down the side of the dish into the hypotonic solution.

(vi)   Immediately remove this hypotonic/fixative mixture and replace with fresh fixative. The dish may be left in the refrigerator at this stage, if necessary. Indeed, this will improve chromosome spreading if the ambient temperature is cool (e.g., in winter).

(vii)   Remove the fixative and replace with more fixative.

(viii) Remove the coverslip from the dish by levering it up with a syringe needle and gripping the edge with forceps (caution: avoid scratching the surface of the coverslip). Drain the coverslip on a piece of labelled Post-Lip paper and lay it flat to air-dry.

(ix)   Mount the coverslip cell side upwards onto a microscope slide with DPX mountant, and dry it overnight.

Cells left growing in dishes or wells after removal of the coverslips can be maintained and trypsinised if necessary. They can either be sub-cultured into Leighton tubes or flasks, or trypsinised directly onto further coverslips for subsequent harvesting by the *in situ* technique.

Preparations may be stained and banded by any of the techniques outlined in Chapter 3. Suspension-harvested slides normally band better if they are 'aged' by overnight incubation in a 60°C oven. Coverslip preparations do not normally need ageing.

(iii) *Chamber cultures.* Harvesting is essentially the same as for coverslip cultures but reagents are applied and withdrawn through the chamber. After the last fixative wash has been removed, the plastic chamber is removed and the slide can then be stained. The gasket should be removed before the coverslip is applied.

3.2 Cell Types

3.2.1 *Fetal Cells*

The morphology of cells found in amniotic fluid is heterogeneous. It has been postulated that the cells may arise from the amnion, skin and urogenitory, respiratory and alimentary systems (6).

When the cell suspension is first placed in cell culture medium the cells appear 'squamous' in type, rounded and flat-looking as they float in the medium. During growth, cells of different size and morphology are observed. Primarily they develop either as fibroblastic or epithelioid in morphology, the former appearing long and thin with a central nucleus, the latter having a more rounded, flattened appearance. Large, polyploid

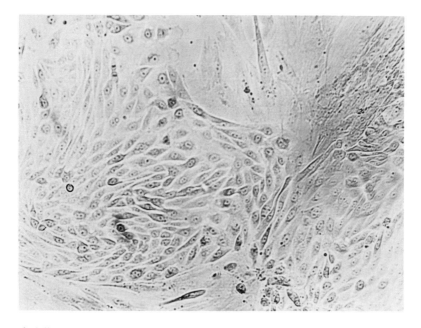

Figure 6. Cell types commonly seen in amniotic fluid cultures.

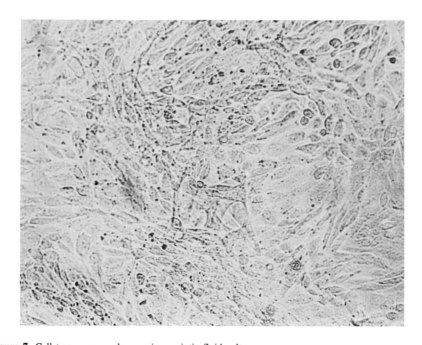

Figure 7. Cell types commonly seen in amniotic fluid cultures.

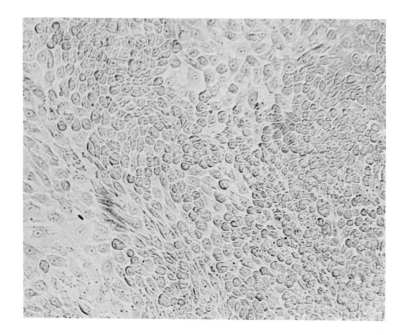

Figure 8. Cell types commonly seen in amniotic fluid cultures.

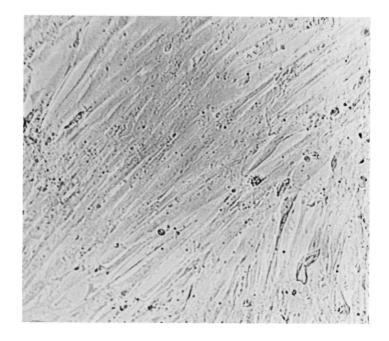

Figure 9. Fibroblasts present in an amniotic fluid culture.

cells are sometimes present, these probably originating from the amnion. *Figures 6−10* illustrate some of the cell types commonly seen in amniotic fluid cultures.

3.2.2 *Maternal Cells*

The extracted amniotic fluid almost always contains maternal blood cells which may give a pink appearance to the samples. A small sample (0.5 ml) of a heavily blood-stained sample should always be taken to a haematology department for a Kleihauer test, to check whether fetal blood cells are present. Maternal macrophages are thought to have a limited survival in culture, but may divide during the first 7 days. Thus, they may present a problem to harvests performed within this time. Since, however, they do not survive for more than 14 days, and any mature macrophages are eliminated by trypsinisation, they are rarely encountered in routine culture.

There are methods available for distinguishing cells of maternal origin if this is necessary, and for this the reader is referred to reference (7). The method of Niazi *et al.* (7) may also be applied to amniotic fluid cultures for tissue typing.

3.3 **Culture Failure**

Most laboratories consider a 98% success rate as reasonable for routine amniotic fluid culture. Failure to obtain a result may occur at any stage, but, given that no technical problems have arisen (e.g., incubator failure, contamination, faulty medium etc.), there are several possible reasons for this.

(i)   No viable cells present in the sample. This becomes more of a problem with late pregnancies (20 weeks or more).

(ii)   Not enough cells in the specimen. This is often a problem with samples taken before 16 weeks gestation.

(iii)   Unsuitable cell type established, for example, cultures composed entirely of large, flattened cells (*Figure 10*) which do not undergo active division, and which may be lost after trypsinisation.

(iv)   Heavily bloodstained sample. This often occurs if the sampling needle has penetrated the placenta. Occasionally, trophoblastic tissue may be present, and it is worthwhile trying to culture this by the method described for chorionic villus culture (Section 4.3.1). Failure may result either because no amniotic fluid cells are present, or because the large number of erythrocytes present prevents amniotic fluid cells from settling on the surface of the vessel. This problem is emphasised in smaller culture vessels such as multiwell dishes.

(v)   Brown sample. This often results from bloodstaining which has occurred earlier in pregnancy and subsequently haemolysed. Samples taken from threatened miscarriages often have this appearance. Such samples are usually full of dead cell debris and may be difficult to grow.

Some slow and difficult cultures may benefit from the substitution of Chang medium for the original culture medium. We have had several cases where this application of Chang medium has produced good results, and other laboratories have also found this.

4. SOLID TISSUE CULTURES

4.1 **Types of Specimen and their Pre-treatments**

All solid tissue samples except chorionic villi should be placed in the wash medium

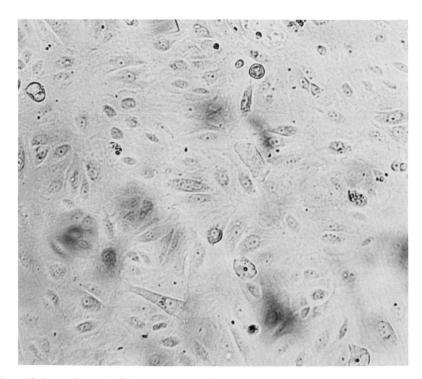

Figure 10. Large, flattened cells in an amniotic fluid culture which do not yield metaphase spreads.

Table 7. Recommended Wash and Transport Medium for Solid Tissues.

Material	Strength	Volume (ml)
RPMI 1640	1 ×	100.0
Fetal calf serum	–	20.0
L-Glutamine	200 mM (100×)	1.0
Penicillin +	10 000 IU/ml +	3.0
Streptomycin	10 000 μg/ml	
Kanamycin	10 000 IU/ml	3.0
Nystatin	10 000 IU/ml	1.5
Sodium bicarbonate	7.5%	1.0
Hepes buffer (if appropriate)	1 M	1.0

described in *Table 7* immediately on receipt and soaked overnight if possible before being set up for culture. Chorionic villus samples should be collected into this medium directly but not left overnight unless a 24-h direct preparation is intended (see Section 4.1.4).

4.1.1 *Skin*

Skin biopsies from live patients are normally performed with a skin punch 1 mm deep and 4 mm diameter, or by tightly pinching a region of skin and cutting off a small sliver. Larger samples of skin are usually available from post-mortem material. As much fat as possible should be removed from the specimen before it is soaked in wash medium.

4.1.2 *Post-mortem Fetuses*

Spontaneously aborted fetuses or those from pregnancies that have been terminated by prostaglandin induction are often received entire. Early pregnancies (i.e., those of 13 weeks gestation or less) are normally terminated by suction, and the specimen received from these may be any part of the fetus or sac. Specimens from early pregnancies are discussed in Section 4.1.3.

When receiving an intact fetus take samples as follows.

(i)   Clean an area of skin with a swab soaked in isopropyl alcohol (e.g., Steret) and remove a piece of about 1 cm². This is best removed from the buttock so that deep muscle may also be removed.

(ii)   Cut deeply into the muscle and remove a small piece of about 1 cm².

(iii)   Make an incision over the sternum, pull back the skin and cut out a small piece of cartilage of about 1 cm².

There is little point in taking samples from fetuses that have been received in formalin since tissue, once preserved, will not grow. There may, however, be a slim chance of establishing a culture from the eyes if the eyelids were closed during immersion in formalin, and if the fetus has not been preserved for more than 1 h or so. Likewise, cultures from fetuses which have been treated with urea *in utero* prior to termination are also unlikely to grow, but, again, it may be worth attempting to culture the eyes.

4.1.3 *Products of Conception*

This term refers to the evacuated contents of the uterus after termination of an early pregnancy, or to the spontaneously aborted material collected after early miscarriage. The nature of the sample may vary from recognisable fetal limbs or ribs and so on, to a large mass of tissue representing parts of the sac, placenta, maternal decidua and blood clots. Many spontaneous abortions are anembryonic, sometimes representing a hydatidiform mole or teratomatous tissue. Products of conception originating from a missed abortion (i.e., a pregnancy which has ceased to develop, but which has not actually aborted) are normally evacuated by suction (hence the term 'Evacuated Retained Products of Conception' or ERPC).

(i)   Fetal parts, and products of conception are normally received in a sterile container and in normal saline. Pour out the contents of the container into a large Petri dish under aseptic conditions. If necessary remove solid material to another Petri dish with forceps and add to it a few drops of wash medium to wash away excess bloodstaining.

(ii)   If recognisable fetal limbs or other parts are present, select these for culture since they will grow more easily than placental, membranous or sac material and are less prone to producing karyotypic artefact. Similarly, if an intact embryo is present amongst tissues from a very early pregnancy, this should be selected for culture, but other tissues should also be taken since the embryo is likely to have died before the sac and may not grow.

(iii) If chorionic villi are present amongst aborted sac material, about 10 mg should be removed for culturing by the method described in Section 4.3.1. If the abortus has been received within 2 − 3 h of suction termination it may be worth attempting a direct preparation as well as a culture (Section 4.3.2).

(iv) Many products of conception will consist of a large mass of placental material or a sac composed of a tough outer skin lined with tissue. It is extremely difficult to know exactly which tissues are which but, as a general rule-of-thumb, it is advisable to avoid any dark red-coloured tissues since these will almost certainly be decidual or maternal blood clot. Pale, fleshy tissue can generally be regarded as fetal and should grow well in culture. Fetal membranes should be avoided if at all possible since these are difficult to grow and are noted for producing karyotype artefacts.

(v) The tissues selected for culture should be soaked in wash medium overnight. If received fresh, they may be stored in wash medium at 4°C for up to 3 days (e.g., over a weekend). It is worth replacing the remainder of the sample into another container of wash medium and keeping this in case the original portions are badly infected. If a culture is clearly infected within 24 h of setting up, it may be worth repeating on the remainer which will, by this time, have been soaked in wash medium for longer, which may be sufficient to inhibit the contaminants.

4.1.4 *Chorionic Villus Samples*

CVS is the most recent prenatal diagnostic technique to be developed. Samples may be taken by a variety of means, the most popular of which are transcervical or transabdominal aspiration. Depending on the skill of the person doing the sampling, chorionic villus samples may be mixed with maternal blood, decidua and cervical mucus. Furthermore, villi may be degenerative if sampled from the chorion laeve (blind aspiration techniques are prone to this). It is important that ultrasound guidance is used so that the vascularised and 'budding' villi of the chorion frondosum are sampled. On receipt, the specimen should be immediately cleaned by removing all solid material from the transport medium and placing it in successive dishes of fresh wash medium until the tissue pieces can be clearly recognised by eye. Villous material can then be separated from other tissues under a dissecting microscope. It is crucial that this step is performed accurately to avoid maternal cell contamination in the culture. Villous material can then be divided up according to the approach to be taken, either long-term culture, direct preparations or both.

4.2 **Culture of Solid Tissues**

4.2.1 *Setting up the Culture*

Most methods of solid tissue culture involve the cutting up of the tissue into small pieces and explanting these into culture vessels. The problem of attaching the tissue pieces to the substrate has been approached in many different ways, including 'glueing' them down with a plasma clot. The method outlined below has produced a very high success rate in our laboratory, and may be used for all types of tissue. Such has been its potential that we were able to adapt it successfully for chorionic villus culture, a particularly difficult tissue to grow under normal conditions. Above all, it is a very simple method.

The medium recommended in *Table 3* is suitable for solid tissue cultures, although some laboratories prefer to use a cheaper medium such as 199, MEM or L-15 supplemented with a lower FCS concentration such as 10%.

Although many samples will grow quite successfully in flasks, we strongly recommend that these are used in parallel with coverslip cultures. Our reasons for this recommendation are 2-fold.

(i)   Samples of less than about 5 mg (or 2 mm²) are very difficult to grow in large vessels. There is a limit to the number of fragments that can be generated by cutting, and many of these will invariably be lost or simply not grow. Plasma clots tend to be inhibitory to the growth of fragments smaller than 1 mm².

The smallest piece of tissue we have grown was a small frond of chorionic villus 2 mm × 1 mm. Four coverslip cultures (13 mm in diameter) eventually provided us with two fully confluent flasks of cells (of male, and therefore fetal karyotype).

(ii)   We have often found that samples manifesting gross microbial infection in flask cultures after overnight incubation have produced sterile coverslip cultures. On occasion, one of four coverslips may eventually become infected, but we have obtained results from the remaining coverslips in many instances (N.B., this may be due to the small well-size of diameter 13 mm. We have not achieved this with larger dishes.)

Thus, for the culture of solid tissues and chorionic villi it is well worth trying an open coverslip culture system.

Chang medium, which has been discussed in Section 2.3.2 (iv), specifically designed for amniotic fluid culture, has been shown to accelerate chorionic villus culture (8,9). Given this information, it would seem only logical to assume that other fetal tissues would do well in it, and this is indeed the case. Fetal limbs and other obvious fetal parts do well in all media, and grow very rapidly. It is not, therefore, worth using Chang medium for these, since other media are considerably cheaper. Chang medium is particularly useful for growing the more obscure tissues such as ERPC material, which, in our experience, are often impossible to grow in other media.

It may well be wondered whether such samples are worth the expense of using Chang medium, given that some people believe that the results of such investigations are of limited value. This depends on the importance placed on such results by the individual clinician and laboratory. If Chang medium is used in the laboratory for prenatal diagnosis work, it is often possible to use left-over aliquots which have been reconstituted and stored for 4 − 7 days, and which are unsuitable for precious cultures. Likewise, we use all our Chang medium which has passed its expiry date for solid tissue cultures, for which it is perfectly suitable.

The following methods apply whether the culture is to be set up in Ham's F10 or Chang medium (*Tables 3* and *8*).

(i)   Remove the tissue from the container of wash medium, with sterile forceps, and place in a plastic Petri dish. Glass Petri dishes are not suitable since the cutting action described below cannot be properly achieved.

(ii)   Take up 0.3 ml of medium into a 1-ml syringe. Remove the needle and drop 0.1 − 0.2 ml medium onto the tissue depending on its size. Rest the syringe with

Table 8. Chang Medium for Chorionic Villus Culture.

Material	Strength	Volume (ml)
Chang B	1×	100
L-Glutamine	200 mM (100×)	1.0
Penicillin+	10 000 IU/ml+	1.0
Streptomycin	10 000 μg/ml	
Nystatin	10 000 IU/ml	0.5

Reconstitute complete Chang medium by adding 2 ml of Chang A or C to 18 ml of Chang B with added supplements as above.

the remaining medium on a nearby object so that the tip does not touch the bench.

(iii) Using a size 22 scalpel blade, make several firm cuts into the tissue. Continue cutting until all tissue has been reduced to a 'mush' of very fine fragments of less than 1 mm². Do not let the tissue dry out during this procedure, if necessary add another drop of medium from the 1-ml syringe. (N.B. Do *not* use scissors for this, it is not possible to achieve a fine enough consistency of the tissue with these, and this fine cutting is the key to the success of this method.) Try not to scrape tissue fragments together with the scalpel blade since fragments of plastic will become mixed with them.

(iv) Add the remaining medium from the 1-ml syringe to the macerate and tilt the dish. Gently scrape tissue fragments into the medium that collects in the crease using the blunt end of the 1-ml syringe.

(v) Take the macerate up into the syringe and inoculate each of four 13 mm coverslips in a multiwell dish with a few tissue fragments (not too many since there will be no room for cell growth).

(vi) Inoculate the remaining macerate onto the surface of a flask and distribute it evenly across the surface by moving the flask and tapping the side. Allow excess medium to run into a bottom corner of the flask, avoiding the tissue fragments. Invert the flask allowing the collected excess medium to run down the crease onto what is now the bottom surface.

(vii) Place a sterile coverslip into each well thus sandwiching the tissue fragments between it and the coverslip onto which they were originally placed.

(viii) Add 5 ml of medium to the flask. Thus the tissue fragments are now attached to what is now the upper surface of the flask and the medium is not in contact with them. The flask is re-inverted on the following morning so that the medium now bathes the adherent tissue fragments. Some pieces will inevitably float into the medium, but there should be plenty adhering. It is therefore essential that this re-inverting should be done *gradually* and *gently*.

(ix) Add about 0.75 ml of medium to each well.

(x) Incubate all cultures at 37°C, with 5% CO_2 and 97% humidity if appropriate to the culture system used (Section 2.5).

This method will work with Hepes-buffered media thus eliminating the need for CO_2, but the coverslip cultures are never quite as quick or of the same quality as with CO_2 buffering.

4.2.2 *Culture Maintenance*

(i)   Leave the cultures undisturbed for at least 5 days. After this they can be ex-
        amined with a ×100 invertoscope. Cells should be seen growing out from the
        cut surfaces of the tissue fragments. Cells on coverslip cultures grow on both
        the upper and the lower coverslips.

(ii)   If cell growth has started, do a complete medium change. Cultures initiated in
        Chang medium must also be renewed. If no cell growth is noted replace in the
        incubator undisturbed.

(iii)   When assessing cultures for harvesting, apply the same criteria as detailed for
        amniotic fluid cultures. Cultures from fetal parts may be ready to harvest after
        only 7 days, whereas adult skins may take up to 3 weeks. Replenish medium
        twice-weekly as with amniotic fluids.

4.2.3 *Harvesting*

This is performed in exactly the same way as for amniotic fluid cultures (Section 3.1.3).
When coverslip cultures are ready for harvesting, the upper coverslip may be removed
by inserting a syringe needle beween the two coverslips of the 'sandwich' and levering
it up so that it can be gripped with forceps. Remember to place the coverslips in their
new well cell-side upwards. Lower coverslips should be harvested at a later date, and
should also be removed to a fresh well 24 h before harvesting.

Staining and banding techniques as described in Chapter 3 may be applied to these
preparations.

4.2.4 *Culture Failure*

The most common reasons for failure of the above solid tissue cultures are, firstly,
contamination of the specimen at source, and secondly, non-viability of post-mortem
specimen. In addition, such cultures may fail if tissue fragments fail to adhere to the
culture vessel and if the sample was not cut sufficiently small. Skin biopsies may fail
to grow if fat has not been properly removed.

4.3 The Use of Chorionic Villus Samples for Prenatal Diagnosis

4.3.1 *Long-term Culture*

As already mentioned, there is no difference between culturing chorionic villus biopsies and culturing other solid tissues. There are, however, a few points which must
be made when considering CVS culture.

(i)   Maternal cell contamination may lead to a mixed maternal/fetal cell culture or,
        in some cases, maternal cells may overgrow. It is therefore important that the
        villous material is properly prepared for culture by careful inspection and removal
        of all maternal tissue. The potential risk of maternal cell contamination may lead
        to serious consequences in prenatal diagnosis and for this reason we recommend
        that laboratories who as yet have no experience with this tissue do not attempt
        routine CVS culture for diagnosis until they have cultured at least 50 practice
        specimens without evidence of maternal cell contamination.

(ii)   There are several reports in the literature of the use of enzyme pre-treatments for CVS culture (11). Although this does, indeed, produce fast results, we suspect that the risk of maternal cell contamination is increased with such methods and we do not recommend them.

(iii) It has been shown (10) that it is the mesenchyme core of the villus that produces most of the cell growth and it is therefore important that the outer trophoblast layers are removed. This is achieved by the fine cutting of the sample. It is also important that this is a cutting action and not a tearing action since cells tend not to grow from torn surfaces.

(iv) We do not recommend that CVS cultures are harvested by the 'suspension harvest' method used for other cultures [Section 3.1.3 (i)]. The following method is better suited to cultures which may have been initiated from a small amount of tissue or which are growing slowly. The principle of this method is that a few cells are removed from the culture leaving plenty behind to re-establish to ensure that the original cell line is maintained. We also use this method for our amniotic fluid and solid tissue harvests, together with coverslip cultures.

CVS Harvesting method

(i) Trypsinise flask and Leighton tube cultures as described in Section 3.1.3 (i) from steps (iii) to (vii) using 0.5 ml of trypsin for both types of vessel, but do not remove all cells from the surface of the flask. This step should only take about 15 sec.

(ii) Neutralise the trypsin with 1 ml of complete medium.

(iii) Using a 1-ml syringe, wash the cell suspension across the surface of the flask to remove more cells. Draw 1 ml of the cell suspension into this syringe.

(iv) Carefully eject 0.5 ml of the cell suspension onto each of two coverslips (22 mm²) in Petri dishes, confining it to the coverslip only. If overspill occurs, 1 ml of medium should be added and the cell suspension spread evenly across the dish.

(iv) Incubate the culture at 37°C. If humidification is used, the coverslips should be incubated overnight. If no humidification is used, the dishes should be topped up with 2 ml of medium once the cells have adhered to the coverslips (usually after about 4 h).

(v) Top up the wells with 2 ml of medium if this has not already been done. Add colcemid (final concentration 0.04 µg/ml) to each dish and incubate for 4 h at 37°C.

(vi) Harvest as for *in situ* coverslip cultures [Section 3.1.3 (ii)].

CVS cultures may fail for a number of reasons. The quality of the sample is very important, and villi sampled from outside the chorion frondosum may be degenerative and therefore not viable. If the *in situ* coverslip method is used then samples as small as 0.5 mg can be successfully cultured (as long as the villi are of good quality). If larger vessels must be used then the minimum sample size should be about 5 mg. The other main reason for failure is insufficient cutting of the sample, therefore attention should be paid to this when setting up.

4.3.2 *The Direct Method*

Although not strictly a culture method, it is necessary to describe the direct method of obtaining chromosome preparations from chorionic villi. The most popular method is that of Simoni *et al.* (8), although other techniques are available (11,12). Direct preparations should always be performed in parallel with a culture from a CVS sample if possible. If there is only a limited amount of villous material available it is up to the individual laboratory and clinician to decide which approach is necessary.

(i)   Wash the villi in 3 ml of Hanks' solution and place them in a Petri dish containing 3 ml of RPMI without serum.

(ii)   Add colcemid to a final concentration of 0.04 μg/ml and leave for 1 h.

(iii)   Remove the medium completely with a Pasteur pipette and replace with 3 ml of 1% sodium citrate solution for 10 min.

(iv)   Remove the citrate completely and add 3 ml of 3:1 methanol/glacial acetic acid fixative. Leave for 10 min.

(v)   Repeat the fixative treatment twice more. Fixed villi can be stored overnight at $-20°C$ at this stage.

(vi)   Remove the fixative thoroughly and gather the villous pieces together in the crease of the Petri dish. Add a few drops (~ 0.5 ml) of freshly made 60% acetic acid solution.

(vii)   Leave the dish tipped for $5-10$ min and observe the dissociation of the cells under phase contrast.

(viii) Prepare the slides by placing a large drop of cell suspension at one end of a cleaned slide placed on a hot-plate at $40°C$ and pulling the drop up and down the slide with the bent tip of a Pasteur pipette until the acid has almost evaporated. Any surplus drops can then be sucked back into the pipette. It is important that the bent pipette tip does not touch the slide during this process, but that it is suspended just above the surface holding the drop by surface tension. Slides can also be prepared using a specially developed machine (GI-RO 3283 multislide machine, Ma-re).

Villi can be incubated for $12-24$ h in complete RPMI medium in a 5% CO_2 atmosphere at $37°C$. The above procedure is then followed.

The results obtained with the direct method tend to be variable in terms of mitotic index and quality of chromosome morphology. The metaphase spreads are extremely difficult to G-band, and for this reason are often Q-banded instead. This type of preparation is also prone to producing broken cells resulting in loss of chromosomes.

4.4 Cell Types

The predominating cell type observed in cultures initiated from solid tissues by the methods described above is the fibroblast. The fibroblast is an undifferentiated cell which grows rapidly in culture. It is also seen in amniotic fluid cultures (*Figure 9*).

Cultures initiated from chorionic villi contain a number of cell types, the predominating one being a spindle-shaped cell probably derived from the mesenchyme core. In addition, there may be small rounded cells and large multi-nucleated cells both thought to derive from the trophoblast layers. Sometimes present in chorionic villus, and also other placental cultures such as those obtained from ERPC material, is a 'convoluted' cell type which is always tetraploid. *Figure 11* shows cells present in a chorionic villus culture.

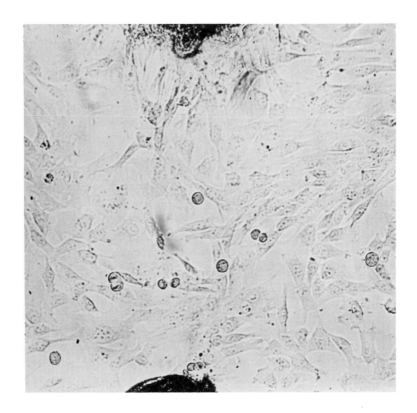

Figure 11. Cells obtained from chorionic villus culture.

Table 9. Medium for Induction of fra(X) in Cell Cultures.

Material	Strength	Volume (ml)
TC 199	1×	94
Fetal calf serum	–	16
L-Glutamine	200 mM (100×)	1
Penicillin+	10 000 IU/ml +	1
Streptomycin	10 000 μg/ml	
Nystatin	10 000 IU/ml	0.5
Sodium bicarbonate	7.5%	1
Hepes buffer (if appropriate)	1 M	1

Supplement the medium with 0.4 μmol FUdR and 60 μg/ml diazepam[a] (final concentrations)

[a]Diazepam is diluted in DMSO to a concentration of 10 μg/ml for the stock solution. It is essential that pure diazepam is used (i.e., not made for injection).

5. TECHNIQUE FOR THE INDUCTION OF FRAGILE X IN CULTURED CELLS

Demonstrating the fragile X [fra(X)] in cultured cells from amniotic fluid and solid tissues has, until recently, been a somewhat difficult task. Consequently, most cases of fra(X) have been detected from lymphocyte cultures. Von Koskull and her co-workers in Helsinki have recently published an improved technique which they have used on cultures of amniotic fluid, fibroblasts and chorionic villi (13).

(i)   Initiate cell cultures and maintain according to the appropriate protocol above. When cultures (passage level 2 − 15) have reached dense confluency remove the medium and replace with medium supplemented with FUdR and diazepam (*Table 9*).

(ii) Incubate at 37°C and inspect after 24 h. Harvest if sufficient mitoses are present. If not, harvest after 48 h.

(iii) Add Colcemid for 2 − 5 min and harvest by trypsin-suspension [Section 3.1.3 (i)].

(iv) Stain slides with 2% Giemsa in buffer (pH 6.8) for 4 min, but do not coverslip. Score for fra(X). Destain and band as required.

6. CULTURE RECORD BOOK

It is essential to keep a careful record of the progress and treatment of all cultures in a log book. The following details should be entered.

(i)   Date of receipt, amount and appearance of specimen.

(ii) Batch numbers of vessels and media used for seeding the culture.

(iii) Number of cultures initiated from the specimen.

(iv) Names of operators at each stage of the duration of the culture (e.g., seeding, medium-changing, harvesting etc).

(vi) Details of progress and treatments carried out (including batch numbers of all media and reagents used).

(vii) Details of any contamination noted.

Such records provide a valuable means of checking any problems such as contamination outbreaks, mixing up of cultures, poor cell growth due to faulty batches of media or reagents and many others.

7. LONG-TERM STORAGE OF CELLS

Cultured cells may be stored indefinitely in liquid nitrogen at temperatures below − 130°C, and many cytogenetic laboratories have a 'cell bank' of precious cell lines, such as those with chromosome abnormalities, which may be used for research purposes.

Freezing, in itself, is lethal to most cells due to damage caused by the formation of ice crystals, and other factors such as denaturation of proteins, dehydration and changes in pH and concentration of electrolytes. The methods for freezing cells are therefore designed to prevent the formation of ice crystals and this may be achieved in two ways. Firstly, cells must be stored at a temperature below − 130°C, and secondly, the cooling rate must be slow enough to allow water to be lost from the cells before they are frozen. The freezing point is lowered by the addition of glycerol or dimethyl sulphoxide (DMSO) to the storage medium.

Cells selected for freezing must be in the logarithmic phase of growth, and the success of the freezing and thawing procedure depends on the age of the cells, the young cells having a greater survival potential. All culture media except Chang medium are suitable for freezing. Complete culture medium is supplemented with 10% glycerol or 14% DMSO and filter-sterilised before use.

7.1 **Procedure for Cryogenic Freezing of Cell Cultures**

(i)   Harvest the cells by the trypsinisation method described in Section 3.1.3 (i) from steps (iii) to (vii).

(ii)   After centrifugation, remove the supernatant and replenish with an equal volume of fresh medium.

(iii)   Remove the supernatant and replace with 3 ml of 'freezing medium'. Mix the cells thoroughly and count in a haemocytometer. Dilute with growth medium to a cell concentration of 5×10^5 viable cells per ml as indicated by trypan blue exclusion [see Chapter 5, Section 2.5.1 (vii) for method].

(iv)   Place 1-ml aliquots of cell suspension into 2-ml freezing ampoules, seal the caps firmly in place, and store in a $-70°C$ freezer for 1 h. Alternatively, if controlled cooling rate apparatus is available, set the controls to achieve a cooling rate of $1-2°C$ per min. When the temperature reaches $-25°C$ the cooling rate can be increased to $5-10°C$ per min.

(v)   Transfer the ampoules to a liquid nitrogen cell storage freezer (e.g., the Union Carbide LR 40) for storage in the vapour phase ($-140°C$) or in the liquid phase ($-190°C$).

7.2 **Thawing Procedure**

(i)   Carefully remove an ampoule from the tray in the liquid nitrogen store and place either in a 37°C water bath or running water from the hot tap. Shake the ampoule continuously until the medium has thawed. It is extremely important that the cells are thawed as rapidly as possible so that ice crystals do not form.

(ii)   Immediately empty the cell suspension into a centrifuge tube that has already been prepared with 8 ml of culture medium. This must be done as soon as the cell suspension has thawed, to minimise the toxic effects of DMSO by diluting it to at least 2%.

(iii)   Centrifuge the tube at 1000 r.p.m. for 10 min.

(iv)   Remove the supernatant and resuspend the pellet in fresh culture medium and seed the vessels as shown in *Table 5*. Cells normally settle within 24 h if the cell density is optimal.

The viability of the recovered cells may be estimated using the trypan blue procedure given in Chapter 5, Section 2.5.1 (vii).

8. ACKNOWLEDGEMENTS

Our thanks are due to the following people for their valuable contributions to this chapter: Oonagh Heron, Lyndal Kearney, Stan Tyms, Richard Moreland and Jackie Marshall.

9. REFERENCES

1. Heaton,D. and Czepulkowski,B. (1984) in *Handbook of Laboratory Health and Safety Measures*, Pal,S.B. (ed.), MTP Press Ltd., Lancaster, p.183.
2. *Code of Practice for the Prevention of Infection in Clinical Laboratories and Post-mortem Rooms* (1978) Her Majesty's Stationery Office, London.
3. Heaton,D.E., Snape,B.M., Fennell,S.J. and Marsden,H.B. (1982) *Prenatal Diagnosis*, **2**, 281.
4. Paul,J. (1975) *Cell and Tissue Culture* (5th edition), published by Church Livingstone Ltd., London.

5. Jakoby,W.B. and Pastan,I.H. eds. (1979) *Methods in Enzymology, Vol.* **58**, published by Academic Press, New York.
6. Van Leeuwen,L., Jacoby,H. and Charles,D. (1965) *Acta Cytol.,* **9**, 442.
7. Niazi,M., Coleman,D.V., Mowbray,J.F. and Blunt,S. (1979) *J. Med. Genet.,* **16**, 21.
8. Simoni,G., Brambati,B., Danesino,C., Rossella,F., Terzoli,G.L., Ferrari,M. and Fraccaro,M. (1983) *Hum. Genet.,* **63**, 349.
9. Heaton,D.E., Czepulkowski,B.H., Horwell,D.H. and Coleman,D.V. (1984) *Prenatal Diagnosis,* **4**, 279.
10. Niazi,M. Coleman,D.V. and Loeffler,F.E. (1981) *Br. J. Obstet. Gynaecol.,* **88**, 1081.
11. Ford,J.H. and Jahnke,A.B. (1983) *Lancet,* **ii**, 1491.
12. Blakemore,K.J., Watson,M.S., Samuelson,J., Breg,W.R. and Mahoney,M.J. (1984) *Am. J. Hum. Genet.,* **36**, 1386.
13. von Koskull,H., Avla,P., Ämmälä,P., Nordström,A.-M., and Rapola,J. (1985) *Hum. Genet.,* **69**, 218.

CHAPTER 2

Lymphocyte Culture for Chromosome Analysis

J.L.WATT and G.S.STEPHEN

1. INTRODUCTION

Peripheral blood is the most easily obtained human tissue, and it will produce a large quantity of good quality mitoses. This has permitted the widespread investigation of the chromosomal basis of many clinical syndromes. The simplicity of obtaining chromosome preparations suitable for analysis from a small volume of peripheral blood has also aided the progress of gene mapping, since family studies are easily performed without causing undue stress to asymptomatic volunteers. There can be little doubt that blood chromosome analysis is the cornerstone of modern clinical cytogenetics.

Reports claiming *in vitro* division of peripheral blood leucocytes date back to 1915, but it was the rediscovery of these mitotically-active cells by Nowell in 1960 (1) combined with the revolutionary air-drying method of Moorhead *et al.* (2) that catalysed the concentration of effort in producing chromosome preparations from white blood cells. It is beyond the scope of this article to give proper historical credit to all pioneers in the development of the techniques described in this chapter, since the main aim is to provide simple instruction, with a view to immediate practical application.

It is inevitable that several of the techniques outlined encroach upon other laboratory disciplines such as haematology, immunology and biochemistry. An attempt to provide minimal background information on these subjects has been made in several places where it is felt to be beneficial.

2. PERIPHERAL BLOOD CELL TYPES

2.1 Basic Haematology

Anti-coagulated blood contains plasma, red cells, white cells and platelets. The red cells (erythrocytes) and the platelets are devoid of nuclei, which renders them totally unsuitable for cytogenetic purposes.

The cytogeneticist's interest therefore centres on the white cells, or leucocytes, which can be divided into five types:

(i)   Neutrophils
(ii)   Eosinophils   } Collectively referred to as granulocytes
(iii)   Basophils
(iv)   Monocytes
(v)   Lymphocytes.

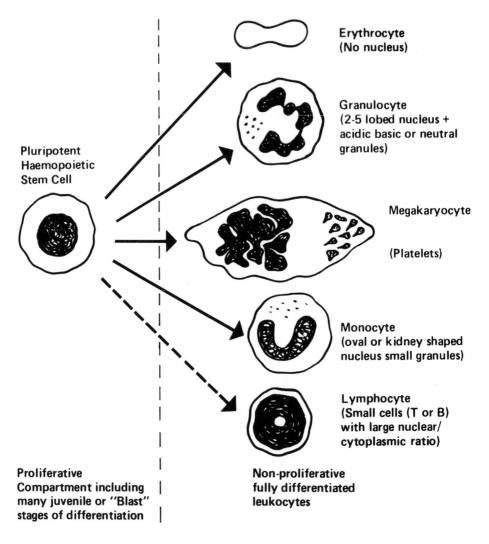

Figure 1. Distinguishing characteristics and developmental origin of leucocytes.

The main distinguishing characteristics and developmental origin of these cell types are summarised in *Figure 1*. In healthy individuals, the fully differentiated forms of all five types of leucocyte in the peripheral circulation do not normally participate in active division, but the lymphocyte can be induced to divide under suitable conditions, and is therefore the cell type of maximum value to the cytogeneticist.

2.2 Sub-sets of Lymphocytes

The lymphocytes are mostly small with a high nuclear/cytoplasmic ratio. There are two major types of lymphocytes, the T cells (thymus dependent) and the B cells (bursa or bone marrow dependent). The former is concerned with cell-mediated immunity, and the latter with humoral immunity. They are not distinguishable on the basis of morphology, but may be identified by surface marker techniques.

Foreign material (antigens) invokes two main types of immune response *in vitro*: the T lymphocytes become sensitised with endogenous surface immunoglobulin, while the B lymphocytes produce free antibody. After exposure to an appropriate antigen, both T and B lymphocytes have the capacity to become large proliferating cells with an immature appearance (blast transformation). T cells are in the majority (70%), and can be divided into Helper, Suppressor and Cytotoxic types. The differentiation of human B cell lymphocytes into antibody-secreting plasma cells is regulated by these functionally and phenotypically distinct sub-sets of T lymphocytes. The identification of the T cell sub-populations has been facilitated by the availability of monoclonal antibodies that recognise various T cell differentiation antigens. Hence the induction of B cell mitotic activity involves T cells, and complicated co-operation between these cell types is essential for immunological competence. It is this natural *in vivo* characteristic of peripheral blood lymphocytes that has been exploited by the cytogeneticist under *in vitro* conditions.

2.3 Mitogens

The success or failure of blood cultures for cytogenetic analysis is dependent on the numbers of normal functioning peripheral lymphocytes at the time of sampling. Therefore, if either the lymphocyte count (particularly the T lymphocytes), or immune competence is reduced, the mitotic index will be low. These critical factors may be radically altered by disease (5) (e.g., Hodgkins disease, leukaemia), infection (e.g., infectious mononucleosis) or drugs (e.g., chemotherapy, contraceptive pill, etc.). Patients who have undergone splenectomy may have a transient, reduced response to PHA, since the spleen is an important storage organ for lymphocytes.

It is now clear that the *in vivo* function of the small circulating lymphocyte is mimicked *in vitro* by exposure to certain antigens, and that these cells become transformed into DNA-synthesising, mitotically active cells in culture. *In vitro* mitotic agents include tuberculin purified protein, tetanus toxoid, diphtheria toxoid, smallpox vaccine, leucocyte antiserum, pollen extract, yeast extract, living or dead donor cells (histocompatibility mismatched), various tissue antigens and plant antigens such as concanavalin A, pokeweed mitogen (PWM) and phytohaemagglutinin (PHA).

2.3.1 *Phytohaemagglutinin*

It was originally believed that PHA-stimulated proliferation is a function attributable to T lymphocytes since it was demonstrated that the lymphocytes from thymectomised irradiated animals do not respond to PHA (3). It is now known that B lymphocytes also respond, but require the help of polyclonal T lymphocytes. However, PHA is a relatively ineffective inducer of T cell-dependent B lymphocytes. The exact mechanism of PHA transformation remains obscure, but it is only necessary during the first 24 h of the induction process [rabbit antiserum to PHA has no effect after 36 h (4)]. PHA is a mucoprotein extracted from the red kidney bean *Phaseolus vulgaris*, which causes an increase in RNA synthesis in cells *in vitro* after a culture lag of 24 h. During the next 24 h, the nucleus enlarges and DNA synthesis begins. The first mitoses are seen around 48 h, with waves at 24-h intervals thereafter. Thus, cultures are normally harvested at 48, 72 or 96 h although divisions will be seen at any time after 48 h. In the initial 24 h, the lymphocytes produce a substance called interleukin-2 (IL-2), also known as lymphocyte growth factor, which perpetuates the mitotic process in a chain reaction manner.

2.3.2 *Pokeweed Mitogen*

PWM is a also a polyclonal T cell activator, but is considerably better than PHA at inducing dependent B lymphocyte transformation. The difference between PHA and PWM lies entirely in the fact that they initially stimulate different sub-populations or clones (identified by monoclonal antibodies) of T lymphocytes (the variation in activation of suppressor T cells accounts for most of the difference).

3. TYPES OF CELL CULTURE FROM PERIPHERAL BLOOD

3.1 **Whole Blood Microculture**

This is the simplest, most widely used method of cell culture for routine cytogenetic purposes. There appears to be little advantage in separating white blood cells (WBC) except, perhaps, with difficult cord or stillbirth blood, or in certain disease situations. The erythrocytes and other blood cell elements seldom interfere with division of lymphocytes in culture, although occasionally excess red blood cells (RBC) can be detrimental since they metabolise vital nutrients in the medium. This problem has been overcome by using a large volume of medium and a relatively small volume of blood (hence the term 'microculture'). Many of the granulocytes degenerate after 48 h, so blood microcultures established from whole blood stored in the refrigerator do not usually contain these cells. If it is necessary to selectively remove them from fresh blood for any reason, they will readily ingest iron particles, since they are phagocytic cells, and can be magnetically removed.

3.2 **24 h Culture of Whole Blood**

This will detect spontaneous blast cells in the circulation from precursors of any of the five leucocyte types, or from erythroblasts. Mitogens are not used for this. This is useful for the study of leukaemias (Chapter 5).

3.3 **Culture of Separated Blood**

3.3.1 *Buffy Layer Culture*

Leucocytes can be separated by centrifugation (115 *g*) and will settle out with the platelets as a pale layer on top of the heavier erythrocytes, and below the plasma supernatant. These can be carefully removed with a wide bore syringe needle or a sterile pipette.

3.3.2 *Gravity Sedimentation*

This performs the same purpose as above, although the leucocytes remain in the plasma above the red cell pellet after standing for 40 − 80 min. However, this varies with temperature and the surface tension of the container. Less than 50% of WBCs from whole blood may be obtained by this method.

3.3.3 *Cell Separation Medium*

Special separation medium such as Ficoll hypaque can be used to separate WBCs more effectively, except in cases where there are few lymphocytes.

4. SAMPLE COLLECTION AND STORAGE

4.1 **Sources of Blood**

Vene-puncture is most commonly used in children and adults, although capillary blood is frequently taken from young babies. The latter may be of a sub-standard quality in cases of heel- or finger-stab where undue pressure has been applied, or bleeding time has been prolonged. Stillborn infants are often difficult to bleed since the veins may have collapsed or the blood may have clotted, and it is sometimes necessary to perform a heart-stab in order to obtain a sufficient volume of blood. The vessels in the umbilical cord are also a suitable source of fetal blood, although this may have deteriorated if not sampled very soon after parturition. Placental blood is less suitable because of the risk of maternal contamination. The source and condition of the sample should always be noted prior to attempting culture.

4.2 **Containers and Anti-coagulants**

Sterile glass or plastic containers, including vacutainers, are equally suitable, and sodium or lithium heparin (preferably without preservatives) is the anti-coagulant of choice. It may be added (0.1 ml of 5000 IU/ml) to the sampling syringe, or, more commonly, tubes may be obtained commercially which are coated with heparin (LIP Services Ltd.). It is possible to obtain chromosome preparations from samples arriving in incorrect containers, even those lacking in anti-coagulant, although the preparations are seldom ideal. However, it is frequently necessary to attempt such cultures, particularly if the sample is a precious one (e.g., from a neonate or stillbirth).

5. WHOLE BLOOD MICROCULTURE

5.1 **Size of Cell Inoculum**

The size of the cell inoculum must be gauged to the individual specimen and the final culture requirement, for instance whether it is to be harvested at 48 h (closer to *in vivo* condition and useful for the study of radiation effects and ataxia telangiectasia) or cultured for a longer period. *Table 1* gives an indication of the variation in differential white

Table 1. Differential Leucocyte Counts for Infants (1−3 days), Children (<6 years) and Adults.

Cells x 10^9	Infants	Children	Adults
Neutrophils	1.5 −13.0	2.0 −6.0	2.0 −7.5
Lymphocytes	2.0 − 8.5	5.5 −8.5	1.5 −4.0
Monocytes	0.3 − 1.5	0.7 −1.5	0.2 −0.8
Eosinophils	0.1 − 2.5	0.3 −0.8	0.04−0.4
Basophils	<0.01− 0.1	<0.01−0.1	<0.01−0.1

Table 2. Recommended Blood/Medium Ratios.

	Blood volume (ml)	*Medium volume (ml)*
Normal adults	0.8	10
Infants and young children	0.6	10
Pregnancy and immediate post-parturition	1.0	10

cell counts (WCC) with age, and may be a useful guide in assessing the initial seeding volume.

Generally speaking, lymphocytes are the predominant cells (up to 60%) until about the sixth year, when neutrophils begin to predominate. Factors affecting differential WCC include age, sex, diurnal rhythm, strenuous exercise, emotional state, smoking, drugs and irradiation. During pregnancy the WCC increases, but this is mostly due to neutrophils, the actual lymphocyte count decreasing in the first trimester and remaining low throughout. Differential WCC is subject to geographic variation, (e.g., the neutrophil/lymphocyte ratio is reversed in tropical areas). *Table 2* gives recommended blood/medium ratios for whole blood microculture.

5.2 **Problem Bloods**

Many of the reasons for failure of standard blood microculture have already been indicated in previous sections. The source of the blood, the container characteristics and the particular circumstances surrounding the donor may all contribute to sub-standard preparations. In the case of stillborn infants, the cells may be dead on arrival at the laboratory. This is usually evident from the initial red cell lysis and the debris from ruptured macerated white cells at the end of culture. Other categories of samples that are frequently problematic include leukaemia, related blood disorders and infection. It is wise to ask for a repeat sample when the patient has fully recovered. Cord, cardiac and capillary bloods show great inter-sample variation. An attempt should be made to establish more than one culture in different media from the fresh sample and, if possible, they should be harvested on alternate days.

Blood from women following abortion or stillbirth is often poor to respond in culture. This is probably influenced by stress and drugs, and it may be advisable to delay taking samples from these patients until some time after the event.

In the case of disease or drug influence, and where the sample is precious, it may be worth attempting one of the following.

(i)   Remove the cells from the influence of their own plasma by washing them with medium or balanced salt solution, after centrifugation.

(ii) Concentrate the white cells from a large volume of whole blood by gravity sedimentation, centrifugation or by using special separating medium (see Chapter 5, Section 2.4.3 for method). Serum batches or donor plasma known to give an improved mitotic index could also be reserved for dealing with problem bloods.

5.3 **Culture Conditions**

5.3.1 *Timing*

The principle of whole blood microculture and the inherent lag period prior to initiation of mitotic action of the lymphocytes has already been described. While there may be individual variation in this 24-h delay, the lymphocyte cell cycle usually begins around this time and takes approximately 24 h to complete. The average length of the DNA synthesis (S) phase in human lymphocytes is approximately 9 h, and G2 is about 3 h (6). The actual mitosis (prophase, metaphase, anaphase and telophase), during which time the chromosomes are visible, lasts only about 1.5 h. Traditional cultures are not synchronised in any way, so that the use of metaphase arresting agents is vital.

PHA-stimulated cultures also have a finite reproductive life-span. With normal initial inocula such as those suggested in *Table 2*, and without changing the medium, the plateau of mitotic activity is reached by the third day, with degeneration at 9 days. However, where the initial cell inoculum has been very low (0.1 − <0.2 ml/10 ml medium), the cultures are most active between 6 and 10 days. The limit of activity is around 14 days, but long-term lymphoblastoid cell lines can be established under appropriate conditions as described in Section 9. For routine purposes, in non-synchronised cultures, 72 h is the recommended total duration of culture.

5.3.2 *Medium*

(i) *Components*. Most of the media described in Chapter 1 (Section 2.3) are suitable for the culture of lymphocytes, including Minimal Essential Medium, Medium 199, Nutrient Mixture F10 and, in particular, RPMI 1640. These media require supplementation with 10−20% serum, either fetal calf, newborn calf or calf sera. Alternatively, human AB serum may be used; indeed, this has been reported as superior to other groups. New batches of serum should always be tested for their ability to support a good mitotic index since this can vary enormously. Once a satisfactory batch has been identified, it is often possible to reserve quantitites of the same batch with the suppliers.

The chemically defined serum substitute, Ultroser G, may also prove suitable for lymphocyte culture, but this has not yet been established. It is possible that, like amniotic fluid culture, some serum supplementation of the order of 2−4% may be necessary.

The reconstitution of media, and the storage of its various components are discussed fully in Chapter 1 (Section 2.3) and readers are referred to this chapter since the information is also applicable to blood culture. The only additional factor necessary for the culture of lymphocytes is a mitogen, normally PHA, as discussed earlier.

PHA (M form) is a crude extract of beans that have been homogenised at 4°C in a 1 M saline solution, and this is normally supplied lyophilised. It should be stored at 4−8°C.

(ii) *Synthetic media*. Modern synthetic growth media have recently become available which require no serum supplement, and which are totally chemically defined. These are considerably more expensive than the traditional medium/serum mixture, but have the advantage of eliminating batch variation, and they appear to produce superior results in terms of mitotic index. An example of a synthetic medium is Chang Medium which is used for the culture of amniotic fluid, chorionic villi, and, more recently, bone marrow, but this has not been widely used for lymphocyte culture. Another synthetic medium, Iscoves modified Dulbecco's Minimal Essential Medium, has been specifically designed for lymphocyte culture without the need for serum, which is replaced by three supplements; bovine serum albumin, human transferrin and soybean lipid.

The most recent modification of this medium is a folate-free formulation, suitable for routine lymphocyte culture, which will highlight fragile sites, in particular the fragile X.

Iscoves media should be stored in the dark at 4°C for up to 2 weeks or at -20°C for longer periods. The supplements should also be stored in the dark at $2-8$°C and may be used until the expiry date. Medium to which supplements have been added can be stored at 4°C for up to 10 days. It is important *not* to freeze either supplements or the complete medium (which in the latter case destroys the effectiveness of the lipid suspension). Supplements should not be used after the expiry date and should be discarded if precipitation occurs or the liquid becomes turbid.

(iii) *Control of contamination.* Since media, sera, PHA, L-glutamine and colcemid are all supplied reliably sterile, and lymphocyte culture is of short duration, contamination is not a prime consideration, providing that good laboratory practice and reasonable aseptic techniques are observed when dispensing solutions and handling cultures. The identification and control of microorganisms is considered in Chapter 1 (Section 2.1.3). The addition of anti-mycotic agents to the medium is not recommended for routine lymphocyte culture, indeed, they should not be necessary since fungal contamination is seldom encountered except in long-term tissue culture. Likewise, infection with mycoplasma is rare but can be inhibited by the use of gentamicin or kanamycin. Bacterial infection can occur, but can generally be adequately controlled by a mixture of penicillin (for Gram-positive bacteria) and streptomycin (for Gram-negative bacteria), or by gentamicin which controls both and has some inhibitory effect on mycoplasma.

5.3.3 *pH and Temperature*

The requirements for temperature mimic the *in vivo* situation in that body temperature of 37°C is ideal, but within the range of $36-38$°C is tolerated. Temperatures of more than 39°C result in death of the culture. Lowering the temperature temporarily (e.g., electric or incubator fault) will not kill the cells, but will be detrimental to preparation quality, and will slow the progress of the cultures so that an adequate mitotic index may not be present at the usual harvest time (72 h).

Similarly, the pH should reflect the near neutral pH of the body fluids, with a range of $7.0-7.4$ being established initially (except for fragile site detection $-$ Section 8). Since lymphocyte cultures are of a comparatively short duration compared with other types of tissue culture, it is sufficient to tightly cap the tubes in an ordinary incubator since endogenously produced CO_2 will serve to buffer the system for a limited time. Some laboratories advocate gassing of the culture vessel with CO_2, or growing lymphocytes in a semi-open system with a 5% CO_2 in air mixture, but such expensive procedures, in our experience, are simply not necessary for routine short-duration blood culturing.

5.4 **Protocol for Lymphocyte Culture**

Tables 3 and *4* detail recommended complete media suitable for lymphocyte culture;

Table 3. Recommended Complete Medium for Culture of Lymphocytes — RPMI.

Material	Strength	Volume (ml)
RPMI 1640	1 ×	100.0
Calf serum	—	20.0
L-Glutamine	200 mM (100×)	1.0
Penicillin/streptomycin	10 000 IU/ml + 10 000 μg/ml	1.0
Sodium bicarbonate	7.5%	1.0
PHA	—	1.5

Dispense 10 ml aliquots of the above into sterile universal containers under aseptic conditions. One bottle of the above is sufficient for 12 individual blood cultures and will keep for up to 2 weeks in the refrigerator, although it is normally used within a few days in a busy laboratory.

Table 4. Recommended Complete Medium for Lymphocyte Culture — Iscoves.

Material	Strength	Volume (ml)
Iscoves medium	—	100.0
Bovine serum albumin	—	1.0
Human transferrin	—	1.0
Soybean lipid	—	2.5
Penicillin/streptomycin	10 000 IU/ml + 10 000 μg/ml	1.0
PHA	—	1.5

Dispense 10 ml aliquots into sterile universal containers and store in the dark at 4°C.

a 'traditional' RMPI recipe and a 'synthetic' Iscoves formulation, respectively.

(i)   Add 0.8 ml of whole blood (or as suggested in *Table 1*) to each 10 ml of medium *immediately* before incubation.

(ii)   Incubate at 37°C for 48, 72 or 96 h as required.

6. HARVESTING TECHNIQUES

6.1 Spindle Inhibitors

As described in the sections on timing, it is necessary to use a mitotic spindle inhibitor, particularly for non-synchronised cultures, so that a block is introduced between metaphase and anaphase. Various chemicals prevent formation of spindle fibres which attach to the centromere of each chromosome prior to mitotic division and pull the chromatids apart by contraction at anaphase. These chemicals include vinblastine, colchicine, and its analogue colcemid. Tijo and Levan (7) adapted the technique from plant cytogenetics.

Colcemid is the Ciba-Geigy trade name for deacetylmethylcolchicine. It is supplied in a liquid form in Hanks' balanced salt solution or in a lyophilised form prepared in Dulbecco's phosphate-buffered saline (PBS). Both have a concentration of 10 μg/ml. The time in colcemid is proportional to the metaphase index, but prolonged exposure results in reduction of chromosome length, since metaphase chromosomes become progressively shorter with time. This phenomenon occurs independently from contraction of the spindle fibres. A compromise situation must therefore be sought where there are adequate numbers of metaphases with chromosomes of a sufficiently elongated morphology to facilitate good resolution banding. This 'compromise' is estimated at $1.5-2$ h colcemid treatment for non-synchronised cultures and a much shorter period, normally $10-15$ min, where cultures are synchronised.

6.2 **Hypotonic Solution**

The history of hypotonic pre-treatment for chromosome preparation dates back to Hsu's fortuitous observations in 1952 (8). A hypotonic solution is one with a lower salt concentration than that in the cytoplasm of cells, so that the rules of osmosis dictate that there will be a movement of H_2O from the hypotonic solution into the cell through the cell membrane. This swells the cells and bursts many of the erythrocytes. Hypotonic treatment is therefore a critical stage, requiring accurate timing, and the procedure serves to complement colcemid treatment in aiding dispersion of the chromosomes. Suitable hypotonic solutions include diluted serum (20% in distilled H_2O), dilute balanced salt solutions, 1% sodium citrate and potassium chloride solution (KCl) normally used at 0.075 M strength. KCl is in common use (9) and is generally accepted as being least damaging to chromosome substructure and related banding suitability.

6.3 **Fixation**

The usual fixative for chromosome preparations is a freshly made mixture of 3 parts alcohol (absolute ethanol or methanol) to 1 part glacial acetic acid. Both the alcohol and the acid should be kept in air-tight containers to prevent water being absorbed. Alcohol with a water content does not hydrate cells well, and the volatile alcohol plays a very important role in flattening and spreading the preparations during air drying. Chromosome morphology is adversely affected by changing the 3:1 ratio, and excess exposure to acetic acid results in premature rupture of the cell membrane with loss of chromosomes.

Whole blood cultures need to be agitated during the initial fixation stage but, in the case of pure leucocyte cultures, the fix can be layered onto the cell pellet and mixing can be done some time later. In whole blood microcultures, haemolysis of the remaining red cells occurs at first fix, when red haemoglobin is visibly converted to dark brown haematin. Lack of adequate and careful mixing results in solid brown lumps in the culture and ultimately gives dirty preparations with a low mitotic index. Should any lumps appear, they should be removed with a pipette and up to six fixations may be necessary to clear the metaphase cells of coating substances. But timing is not critical after the first fix, providing the cells are in freshly prepared fix for slide preparation. Insufficient fixation interferes with the flattening of the preparations on slides, and coating substances interfere with banding. Evidence of poor fixation includes poorly spread, 'three dimensional' preparations with a 'glazed' appearance under phase contrast microscopy, and/or chromosomes with an opaque 'halo' of amorphous material giving a 'fried-egg' impression, so that in subsequent staining the chromosomes lose clarity due to the background material picking up stain.

6.4 Slide Preparation and Staining

Laboratories vary in their preparation of microscope slides. Some use slides straight from the manufacturers box, some soak slides in alcohol, fix, ether or chromic acid, and dry and polish slides prior to use. It used to be our practice to pre-treat slides with a detergent solution to remove all traces of grease, but we found that the detergent itself formed a 'coating layer' that took up stain and introduced background colour to the stained preparations. Whether pre-treated for extra cleanliness or not, slides may be washed in distilled water and used wet, or polished and used dry, depending on personal preference. Suitable stains include orcein, Giemsa or other Romanowsky stains. A full range of staining and banding methods is included in Chapter 3.

6.5 Suggested Harvesting Protocol

(i)   Remove the culture vessels from the incubator 2 h prior to harvesting (e.g., at 70 h).

(ii)   Add colcemid (0.2 ml stock solution/10 ml culture) to give a final concentration of approximately 0.2 μg/ml (much less is required for synchronised cultures).

(iii)   Mix gently and return to the 37°C incubator.

(iv)   Remove the culture at 72 h and split the contents between two 10 ml plastic centrifuge tubes, labelled with culture details (patient referral numbers, etc.) *in pencil* (fixative removes ink).

(v)   Centrifuge at 130 *g* for 10 min.

(vi)   Remove and discard the supernatant with a pipette (which should be labelled to match the tube) or with an Edwards' water suction pump.

(vii)   Add 10 ml of hypotonic 0.075 M KCl, pre-warmed to 37°C in a water bath. Agitate using a Pasteur pipette or whirlimixer. The pipette agitation should be minimal and involve the stem of the pipette only since cells are lost in the bulb of pipettes.

(viii) Incubate at 37°C in a water bath for 2 – 10 min depending on the slide-making method used (shorter hypotonic treatments are normally used with 'dropping' techniques).

(ix)   Spin at 130 *g* for 9 min.

(x)   Remove the supernatant and *slowly* add 10 ml of cold, freshly prepared fixative (3 parts methanol:1 part glacial acetic acid). This is best done on a whirlimixer, adding the fix almost dropwise at first, followed by a slow trickle. This is also a very critical stage and every effort should be made to avoid lumps. If formed, these should be removed *prior* to the next centrifugation.

(xi)   Centrifuge at 130 *g* for 10 min.

Steps (x) and (xi) should be repeated until the cell pellet is white and the fix is clear (usually twice more).

(xii)   Remove the supernatant leaving approximately 0.5 ml above the cell pellet. Mix on a whirlimixer to form a milky fluid.

(xiii) Drop at arms length onto shaken, wet, cold, grease-free slides. Usually three drops are adequate.

(xiv) Examine the preparations under phase contrast and dilute or concentrate the cell suspension as necessary for the remaining slides. Eight to ten slides are normally made from each culture.

7. SYNCHRONISATION OF CULTURES

There have been several recent modifications to the standard lymphocyte culture method. These are aimed at allowing analysis of human chromosomes at earlier stages of cell division in order to gain additional information from more elongated chromosomes which permit high resolution banding. The limiting factor in standard non-synchronised culture is the fact that the proportion of early metaphase and prometaphase cells is low compared with mid- and late-metaphase cells, because of the nature of the colcemid arrest. Arresting agents which specifically stop prophase and prometaphase have not been identified, but it is possible to introduce a chemical block at an earlier stage of the cell cycle, so that when cultures are subsequently released from the block, the cells proceed in synchrony to complete division. When this reasoning is combined with knowledge of the cell cycle, careful timing allows the harvest of cultures with a high proportion of prophase, prometaphase or early metaphase cells as required. A short exposure to colcemid is recommended.

7.1 **Methotrexate**

One such chemical-blocking agent is amethopterin or methotrexate (MTX) which is an analogue of folic acid (FA) with a higher affinity for dihydrofolate reductase than FA, so that the synthesis of folinic acid is potently inhibited. Folinic acid is required for the production of thymidine which in turn is required for DNA synthesis. The cells are therefore blocked prior to DNA synthesis at the G1/S interface (see diagram in Section 8.2).

The MTX block can be released by washing (optional) and adding thymidine. The optimum period between release of the block and harvesting must be precisely determined. This period has been variously reported (5−6 h), but Yunis *et al.* (10) state that 5 h 10 min is optimal. Minimum exposure of colcemid is important for the inhibition of spindle formation while having a negligible effect on chromosome condensation. This technique, and others of its kind, is beginning to find widespread use in clinical cytogenetics and a few laboratories have modified it for routine use. The thymidine analogue bromodeoxyuridine (BrdU) can be effectively substituted for thymidine to release the MTX block which normally lasts several hours (e.g., 17 h). BrdU has the added advantage of mildly inhibiting chromosome condensation.

7.2 **Alternative Blocking Agents**

It is perhaps surprising that very high concentrations of thymidine (the MTX block release) can also cause a partial synchronisation block by feedback inhibition from the TTP pool (11). Similarly fluorodeoxyuridine (FUdR) is an antagonist of the enzyme thymidinylate synthetase and can also be used as a G1/S blocking agent (12). Anti-condensing agents include acridine orange, 33258 Hoechst, actinomycin D and ethidium bromide. These have a more marked effect on restricting chromosome contraction and are useful when added during the final period of culture.

7.3 **Practical Considerations**

The original MTX block method (13) is rather long and laborious, and not compatible with routine laboratory work since multiple medium changes and excessive fixation elevate time and costs. Furthermore, many laboratories have experienced difficulties in establishing a reliable and reproducible method. These difficulties are likely to relate to variation in lymphocyte cell cycle under non-constant culture conditions. Different serum batches may influence cell cycle time so that the block release time must be tailored to the particular conditions in an individual laboratory. Slide-making technique is also very important since the long chromosomes must be given room to spread, and membrane bursting is frequently encountered. Some laboratories have devised elaborate plumb-line systems with tilted slides at precise angles to receive a single drop from a height of up to 1 metre! Fortunately, synchronised cultures tend to have a high mitotic index, so that some cell loss can be tolerated.

Since the overnight MTX or thymidine block effectively introduces another day in culture, and the whole procedure in a diagnostic laboratory must be geared towards harvesting convenience, some laboratories now start incubation on 2 days only (e.g., Monday and Saturday) so that harvesting is done on two weekdays, namely Friday and Wednesday, respectively.

7.4 **Suggested Synchronisation Methods**

7.4.1 *Method 1: Methotrexate Block*

(i)   Establish the culture as described in Section 5.4.

(ii)   After 48 h of culture, add methotrexate to a final concentration of 0.45 μg/ml. *This is the 'block introduction stage'.*

(iii) Incubate at 37°C overnight (approximately 17 h).

(iv)   Remove from the incubator and carefully remove the medium from the settled cells.

(v)   Resuspend the lymphocytes in fresh medium containing excess thymidine (0.24 μg/ml final concentration). (BrdU can be substituted for thymidine.) *This is the 'block removal stage'.* It is also possible to simply add excess thymidine (or thymidine analogue) without going to the trouble and expense of changing the medium, although greater quantitites must be added to overcome the block.

(vi)   Incubate at 37°C for a further 5 h 10 min (or the optimum for the laboratory).

(vii) Add dilute colcemid (0.05 μg/ml) for the last 10 − 15 min of culture. (A range of anti-contractile agents *may* also be added.)

(viii) Harvest as described in Section 6.

7.4.2 *Method 2: Thymidine Block*

This method is similar to one which was found to produce outstandingly good preparations by the 1984 ACC/NEQAS Quality Assessment Scheme (14). Some laboratories have found this to be suitable for routine use, since it produces consistently good results without the necessity for the use of dangerous cytotoxic drugs such as methotrexate.

(i)   Establish the culture as described in Section 5.4.

(ii)   After 48 or 72 h of culture add 0.2 ml of thymidine (15 mg/ml in sterile PBS).

(iii)   Incubate at 37°C overnight (14 − 18 h).

(iv)   Remove the old medium from the settled cells.

(v)   Wash in fresh RPMI, centrifuge at 115 g, and remove the supernatant.

(vi)   Replace with 10 ml of complete medium without PHA and incubate at 37°C for a further 5 h.

(vii)   Add colcemid solution (final concentration of 0.1 μg) for 10 min.

(viii) Harvest as described in Section 6.

8. THE DETECTION OF FRAGILE SITES

8.1 Definition of Fragile Site

A fragile site is defined (15) as a specific point on a chromosome which manifests itself (under appropriate *in vitro* conditions) as a non-staining gap of variable width which usually involves both chromatids and may form an acentric fragment or triradial figure. The site is always at exactly the same point on the chromosome in cells examined from any one individual or kindred, and is inherited in Mendelian dominant fashion.

Fragile sites are not seen in 100% of cells and, with the exception of the fragile site which appears at 16q22 in cultures in complete medium, the majority depend for their expression on the culture conditions.

There are three main categories of fragile site:

(i)   those that appear in folic acid- and thymidine-deficient medium (majority, including fragile X);

(ii)   those that appear after exposure to bromodeoxyuridine (e.g., 10q25);

(iii)   a new class of fragile site, sensitive to 5-azacytidine (16).

8.2 Detection of the Fragile X

The culture conditions are crucially important. Medium 199 or Iscove's (folic acid free) are suitable commercially available culture media, although the additives have to be carefully assessed for maximum efficiency. The three most important features of a fragile site medium appear to be deficiency of folic acid, high pH and low serum content. Additives, such as methotrexate, FUdR or methionine may also improve detection.

The underlying biochemistry and the observed nature of inducers and inhibitors of fragile X expression indicate that the process of expression is likely to be operating during DNA synthesis due to a limited dTMP pool. The area of metabolism involved is summarised in *Figure 2*.

Recombinant DNA technology has improved pre-natal detection since four restriction fragment length polymorphisms (RFLP) have been identified on the subtelomeric region Xq28-Xqter (17). There was some controversy over the exact breakpoint involved in fragile X expression with some authors reporting Xq27 and others Xq28. However, a recent high-resolution report (18) has pinpointed it to Xq27.3.

8.3 Suggested Protocol for Fragile X Detection

(i)   Medium 199 without folic acid is a suitable base medium.

(ii)   Add sodium bicarbonate to a final pH of 7.8 − 8.0. The pH at *the end* of the

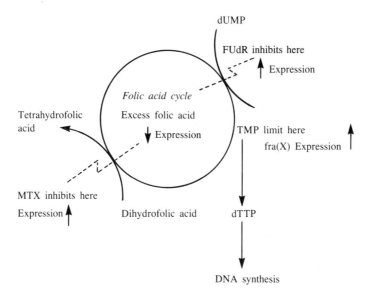

Figure 2. The area of metabolism affected by inducers and inhibitors of fragile X expression.

3-day culture period is actually more critical, so it is worthwhile checking on the second day and altering as necessary.

(iii)   Add as little serum as gives a reasonable mitotic index (some laboratories report growth *without* any serum, but we find this gives a scant mitotic index). Serum addition should not exceed 10%, and is best at between 2 and 5%.

(iv)   (optional) Add FUdR to a final concentration of 5 μM (15).

or

(optional) Add L-methionine to a final concentration 100 mg/l (20).

(v)   Seed 10 ml cultures with 0.8 ml of blood as described previously in Section 5.4.

(vi)   (optional) Add 10 mg/l of methotrexate 24 h prior to harvest (20).

(vii)   Add colcemid at 70 h (0.2 μg/ml). Incubate at 37°C for 2 h.

(viii)   Harvest at 72 h as described previously (Section 6).

Notes

(i) In order to obtain prometaphases, this technique can be combined with a MTX block.

(ii) Solid staining with Giemsa should originally be employed to clearly identify the fragile site (*Figure 3a*) followed by G-banding to verify that the C-group chromosome in question is indeed the X (*Figure 3b*).

9. LYMPHOBLASTOID CELL LINES

These are transformed cultures which can be established from blood of normal donors (21). The establishment may be spontaneous [in cases where there has been previous exposure to Epstein Barr virus (EBV)] (22), or induced by EBV.

Lymphoblastoid cells grow as suspension cultures, and have enormous proliferative capacity, doubling their numbers every 48 h or so. They do not become senescent and can be maintained indefinitely. A successful establishment may take from 50 to more

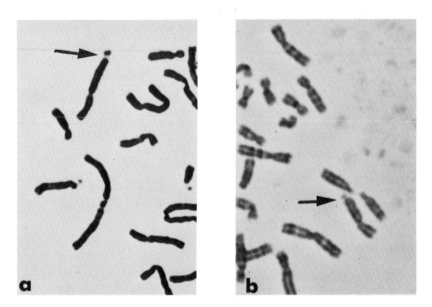

Figure 3. (a) Fragile X seen by solid staining in an early metaphase; (b) identified as X by G-banding.

than 100 days. These cultures can be subcultured and frozen in liquid nitrogen and are a good source of bulk cells without having to resort to the more invasive techniques of skin biopsy. The karyotype is normally stable and reflects the original donor karyotype, so they are suitable for biochemical and gene mapping studies.

9.1 Method of Establishing B-lymphoblastic Cell Line

(i)   Allow heparinised or defibrinated blood (20 ml) to settle by gravitation, and remove the WBCs in the plasma layer, or use a lymphocyte separation product (e.g., Ficoll hypaque).

(ii)   Wash the lymphocytes twice in medium without serum.

(iii) Set up replicate cultures in 25 cm² tissue culture flasks using medium with 20% foetal calf serum (FCS).

(iv)   Add 1 μg/ml of PHA.

(v)   Incubate at 37°C in 5% CO_2.

(vi)   After 48 h remove and centrifuge (130 *g*).

(vii) Resuspend in fresh medium (without PHA) with 20% FCS.

(viii) (Optional in patients with known history of EBV exposure, e.g., infectious mononucleosis, Burkitt's lymphoma, etc.) Add 0.5 ml of EBV virus suspension obtained from an established lymphoblastoid culture (by passing a cell-free suspension through a 0.22 μm Millipore filter).

(ix)   Leave undisturbed for 1 week, then feed once a week by removing half the medium and adding fresh.

(x)   Subculture and harvest as for other tissue cultures (Chapter 1) but trypsinisation is not necessary.

EBV-infected cultures should be handled with care and dealt with in isolation from other types of cell culture. Aseptic technique and the use of antibiotics and antimycotics are recommended (Chapter 1). For further information on long-term lymphoblastoid cell lines, the reader is referred to Glade and Beratis (23).

10. REFERENCES

1. Nowell,P.C. (1960) *Cancer Res.*, **20**, 462.
2. Moorehead,P.S., Nowell,P.C., Mellman,W.J., Battips,D.M. and Hungerford,D.A. (1960) *Exp. Cell Res.*, **20**, 613.
3. Davies,A.J.S., Festenstein,H., Leuchars,E., Wallis,V.J. and Doenhoff,M.J. (1968) *Lancet*, **I**, 183.
4. Younkin,L.H. (1972) *Exp. Cell Res.*, **75**, 1.
5. Cawley,J.C. (1983) *Hematology, Integrated Clinical Science*, published by Heinmann Medical Books Ltd., London.
6. Cave,M.J. (1966) *Cell Biol.*, **29**, 209.
7. Tjio,J.H. and Levan,A. (1956) *Hereditas*, **42**, 1.
8. Hsu,T.C. (1952) *J. Hered.*, **43**, 167.
9. Hungerford,D.A. (1965) *Stain Technol.*, **40**, 333.
10. Yunis,J.J., Sawyer,J.R. and Ball,D.W. (1978) *Chromosoma (Berlin)*, **67**, 293.
11. Schempp,W., Sigwarth,I. and Vogel,W. (1978) *Hum. Genet.*, **45**, 199.
12. Webber,L.M. and Garson,M. (1983) *Cancer Genet. Cytogenet.*, **8**, 123.
13. Yunis,J.J. (1976) *Science (Wash.)*, **191**, 1268.
14. Davison,E.V. (1985) *Clin. Cytogenet. Bull.*, **1**, 168.
15. Sutherland,G.R. (1979) *Am. J. Hum. Genet.*, **31**, 125.
16. Sutherland,G.R., Jacky,P.B. and Baker,E.G. (1984) *Am. J. Hum. Genet.*, **36**, 110.
17. Weiacker,P., Davies,K.E., Cooke,H.J., Pearson,P.L., Williamson,R., Bhattacharya,S., Zummer,J. and Ropers,H.-H. (1984) *Am. J. Hum. Genet.*, **36**, 265.
18. Harrison,C.J., Jack,E.M., Allen,T.D. and Harris,R. (1983) *J. Med. Genet.*, **20**, 280.
19. McDermott,A., Walters,R., Howell,R.T. and Gardner,A. (1983) *J. Med. Genet.*, **20**, 169.
20. de Arce,M.A. and Kearns,A. (1984) *J. Med. Genet.*, **21**, 84.
21. Moore,G.E., Gerner,R.E. and Franklin,H.A. (1967) *Jama*, **199**, 519.
22. Epstein,M.A. and Barr,Y.M. (1964) *Lancet*, **I**, 252.
23. Glade,P.R. and Beratis,N.G. (1976) *Prog. Med. Genet.*, **1**, 1.

CHAPTER 3

Chromosome Staining and Banding Techniques

P.A.BENN and M.A.PERLE

1. INTRODUCTION

To the cytogeneticist, the appearance of well prepared, clearly banded chromosomes have an aesthetic appeal which is often difficult for the non-cytogeneticist to comprehend. In part, this may be attributable to steps in some procedures which have no obvious scientific explanation but which nevertheless do materially affect results. For example, it is difficult to explain why some lots of stain work better than others or why a slide stained horizontally can be so different from one stained in a Coplin jar. Many published staining methods devised in one laboratory require modification in another laboratory. Despite these somewhat mystical aspects of the craft, rigid adherence to times, concentrations, temperatures and pH can result in methods which are highly reproducible and reliable.

The methods described in this chapter represent procedures which the authors have found useful in their laboratories. These protocols are known to be in use in a number of laboratories with minimal variations and refinements. Optimal use of the microscope and good photographic procedures are essential and these issues are therefore briefly reviewed before describing specific banding methods.

2. MICROSCOPY

A top quality research standard microscope is essential for viewing and photographing chromosomes.

An initial familiarisation with the microscope in terms of light path alignment, phase adjustment, maintenance and cleaning is essential. These features are covered in detail in the manuals provided with the microscopes, and reference (1). Good maintenance procedures cannot be sufficiently stressed.

Most chromosome analysis is carried out using transmitted light, 'bright field'. 'Kohler illumination' uses a controlled light path which evenly illuminates the object. Preparations stained with non-fluorescent dyes which show sufficient contrast are generally viewed in this manner. Objectives suitable for this work are 'plan achromats' (i.e., light of different colours are not separated). Generally, $10\times$, $40\times$, $70\times$ and $100\times$ objectives are needed. 'Plan' lenses provide a flat field. 'Neofluar' are high aperture objectives with improved colour rendition and high contrast. High power objectives, 'immersion objectives', require oil (or water in some instances) between the preparation or coverslip and the lens. A particularly useful lens for cytogenetics is the Zeiss 'epiplan' $80\times$ which provides a flat field without the need for oil immersion. This lens has a short working distance between the object and the lens and therefore cannot be used with a coverslip.

Unstained chromosomes and those palely stained with insufficient contrast are viewed with phase contrast illumination. Microscopes are fitted with a phase contrast condenser which must be correctly centred using a focusing 'telescope'. In some microscopes the telescope is built in. Matched to the condenser are special phase contrast objectives.

Fluorescence microscopy is carried out for a variety of chromosome banding techniques (see below). Generally, illumination is achieved by a high pressure mercury lamp with a limited lifespan of approximately 200 h. Xenon and halogen quartz light sources can also be used with appropriate filter combinations. With incident light illumination (as opposed to transmitted light illumination) light passes through an exciting filter and then is reflected down through the microscope objective by means of a 'dichroic beam-splitting mirror'. Light emitted from the object passes back into the lens, passes through the dichroic beam-splitting mirror with unwanted u.v. wavelengths being reflected. A 'supression filter' or 'barrier filter' is usually inserted between the mirror and the eyepieces. The microscope should have the capacity to divert all, or at least 90%, of the image to the camera system to minimise exposure times.

Modern fluorescence microscopes with mercury lamps are sold with filter packages which optimise conditions for particular stains. For the cytogenetics laboratory, filter packages sold for fluorescein isothiocyanate (FITC) dyes are suitable for quinacrine-stained chromosomes. The combination of barrier filter BG 38 (broad band pass above 300 nm with supression of red) or BG 12 (330 – 500 nm), dichroic filter of 510 nm and barrier filter LP 520 are suitable for quinacrine banding, acridine orange, and chromomycin A3 reverse banding. The combination is also suitable for Hoechst 33258 and DAPI-based staining although BG 38 or UG-1 (300 – 400 nm) with TK 400 dichroic and 460 nm barrier are also used. Leitz filter block A has application for quinacrine, Hoechst 33258, DAPI and acridine orange. Filter block E3 can be used for chromomycin A3 reverse banding.

Special objectives are needed for fluorescence with a variable diaphragm required for high power lenses. Low fluorescence oil is also required for immersion lenses.

3. PHOTOGRAPHY

Microscopes with built-in (integrated) camera systems with automatic exposure are usually used in the cytogenetics laboratory. While such systems are expensive, the savings in time and materials are considerable. Integrated cameras also have the advantages of being more stable and having precise alignment of the camera with the image seen through the microscope eyepieces. Most laboratories use a 35 mm format, the 4" × 5" Polaroid systems being expensive in materials.

For light microscopy, both bright field and phase, the 35 mm film increasingly being used is Kodak Technical Pan 2415 film (Kodak catalogue number 129 9916). This film has a very fine grain and provides excellent contrast when developed in Kodak HC-110 developer (140 8998), diluted one part stock with nine parts water for 6 min. Automatic photomicroscopes should be set at approximately 50 ASA (18 DIN) for optimal pictures.

Kodak Technical Pan 2415 film can also be used for fluorescence photomicroscopy using an approximate ASA setting of 500 (DIN 28) for Q-banding, 320 ASA (DIN

26) or 400 ASA (DIN 27) for R-banding by the chromomycin-A3 method, distamycin-Hoechst 33258 staining, replication studies, and so on. However, exposure times may be excessive (>2 min) giving poor results due to fading. For duller fluorescence stains, use Panatomic-X film and set the ASA one stop faster. This film will give good contrast although photographs will have a grainy appearance. Panatomic-X film can be developed in Kodak HC-110 using a dilution of one part stock plus seven parts water for 4.5 min.

Note that film exposure times should, in general, be aimed at the optimum recommended for the particular type of film. Under-exposure with subsequent over-development will give a negative which gives a very grainy appearance with increased contrast ('pushing' the film speed). Conversely, over-exposure and under-development gives a negative which has reduced contrast and this may result in a loss of detail ('pulling' the film speed).

For economy, 35 mm can be purchased in bulk (30 metre or 50 metre lengths) and cut to size, using a bulk film cassette loader. Considerable time can be saved in processing by using a 4% acetic acid stop bath between developer and fixer, Permawash (Heico Whittaker) to reduce film washing time and Kodak Photoflow 200 (Kodak 146 4510) to prevent water marks.

For good pictures of chromosomes there is little point in having an excellent microscope and a poor enlarger. A sturdy condenser type enlarger fitted with good optics (a 50 mm lens for 35 mm work and $135-150$ mm lens for $4'' \times 5''$ negatives) is therefore necessary.

The darkrooms of clinical laboratories preparing large numbers of photomicrographs generally contain an automatic print processing machine, for example the Agfa Rapidoprint DD37E. This machine has four chambers: activator (Agfa G182B catalogue number FW25) and stabiliser (Agfa G382B catalogue FW2V) are placed in the first two chambers. By placing fixer (Agfa QS0 3, catalogue FYA9) and water in the second two chambers, prints can remain in good condition for years after they are made. Processers with only two chambers give pictures that tend to fade or turn brown within a few weeks or months.

Resin-coated photographic paper (e.g. Agfa Rapidtone papers) can be quickly dried. A hot air dryer such as the Arkay RC-1100 can be positioned and the speed adjusted such that prints from the print processing machine feed directly into the dryer. Grades P1-2 and P1-3 are the best for light microscopy and Agfa Rapidoprint TP6 WP is particularly useful for fluorescent banding methods. In general, a series of prints is needed for the fluorescent techniques in order to see the bands on both the brightest and dullest chromosomes.

A useful general book on photographic technique is helpful in becoming familiar with basic techniques in printing and developing as well as in problem solving. The Darkroom Handbook (2) covers basic procedures as well as special techniques including the use of colour.

4. STAINING AND BANDING TECHNIQUES

4.1 Solid Staining

Staining procedures which provide a uniform unbanded appearance to chromosomes are referred to as solid or conventional staining (*Figure 1*). These methods have become

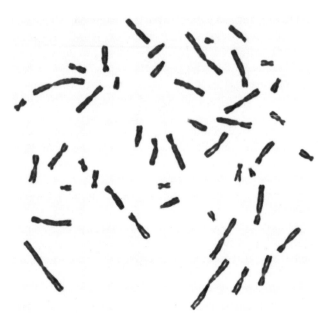

Figure 1. Conventional or solid stained 46,XY normal male cell, Giemsa staining (Section 4.1.2).

largely obsolete since the introduction of chromosome banding. They can, however, be useful for studies on chromosome breakage as scoring gaps and breaks can be difficult in lightly stained chromosome bands.

4.1.1 *Method 1. Aceto-orcein staining*

Particularly sharp, well-defined chromatin staining is achieved with aceto-orcein stain. The chemical basis for the stain reaction is obscure.

(i)   Prepare 2% acetic-orcein by dissolving 40 g of synthetic orcein in 200 ml of glacial acetic acid at 80°C stirring for 4 − 5 h in a fume hood.

(ii)   Adjust the volume to 2 litres with distilled water and filter the solution through three thicknesses of Whatman No. 1 filter paper.

(iii)   Store in a darkened bottle and filter freshly before use.

Prepared 2% aceto-orcein stain can also be purchased.

Procedure.

(i)   Immerse the slides overnight at room temperature in the 2% aceto-orcein stain or 3 h at 37°C. (For preparation suitable for phase microscopy, 10 min staining is sufficient).

(ii)   Rinse twice in 45% glacial acetic acid.

(iii)   Rinse three times in Cellosolve (2-ethoxyethanol).

(iv)   Rinse twice in euparal essence.

(v)   Mount, if necessary, a No. 1 coverslip on the preparation using Clay Adams Histoclad or similar mounting medium.

4.1.2 *Method 2. Giemsa Staining*

This method is much faster than the previous one described although background stain deposit and uniformity of staining is poorer. Giemsa stain is a Romanowsky-type dye mixture. Azure B and eosin in the Giemsa bind to DNA (3).

Prepare the following solutions.

(i)   Phosphate buffer (pH 6.8). Buffer consists of 0.025 M KH_2PO_4 (3.4 g/l) titrated to pH 6.8 with 50% NaOH.

(ii)   Giemsa stain. 5 ml of Giemsa plus 45 ml of phosphate buffer (pH 6.8).

Procedure.

(i)   Place the slides in the Giemsa stain for 8 min.

(ii)   Rinse the slides twice in de-ionised or distilled water.

(iii)   Air-dry (a hair dryer is a useful way to expedite drying).

(iv)   Mount if necessary with a coverslip as above [Section 4.1.1(v)].

Slides can be de-stained by soaking in Carnoy's fixative (three parts absolute methanol and one part glacial acetic acid) and subsequently stained by another technique.

4.2 Giemsa Banding

Giemsa banding (G-banding) has become the most widely used technique for the routine staining of mammalian chromosomes (*Figure 2*). The most usual methods to obtain this staining are to treat slides with a protease such as trypsin (4) or incubate the slides in hot saline-citrate (5), although a variety of other methods have been used. The chromosome banding patterns obtained are thought to reflect both the structural and functional composition of chromosomes (6). Dark bands correlate with pachytene chromomeres, generally replicate their DNA late in S-phase, contain A+T-rich DNA, appear to contain relatively few active genes and may differ from light bands in terms of protein composition (7). Differential extraction of protein during fixation and banding pre-treatments from different regions of the chromosome may be important in the mechanism by which G-bands are obtained (7).

In general, slides for G-banding should be aged 3−5 days at room temperature or overnight (16−18 h) at 56−60°C for optimal results. Slides can be left at 56−60°C for several days without any problems although older slides will require higher trypsin concentrations or other more aggressive treatments to obtain good banding.

4.2.1 *Method 1. G-Banding*

This particular method (8) has been found to show considerable latitude, yielding acceptable results with preparations made by different cytogeneticists using various harvesting techniques.

Prepare the following solutions.

(i)   Giemsa stain solution concentrate. Add 1.0 g Giemsa powder to 66 ml methanol and 66 ml glycerin and stir for 2 days at room temperature. Stain should be prepared at least 2 weeks before use and stored in a darkened container in a refrigerator.

(ii)   Phosphate buffer (pH 6.8) as in Section 4.1.2.

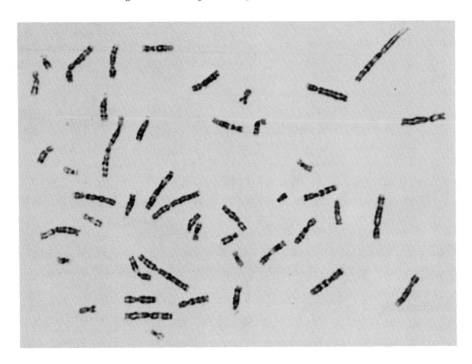

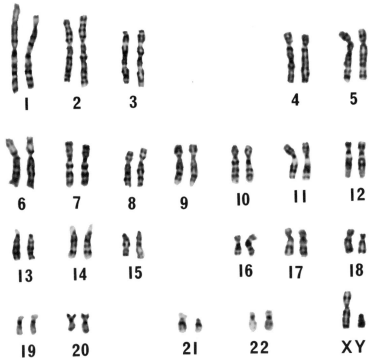

Figure 2a. Giemsa-banded cell (top) with karyotype (below) from a normal male, 46,XY (Section 4.2.1).

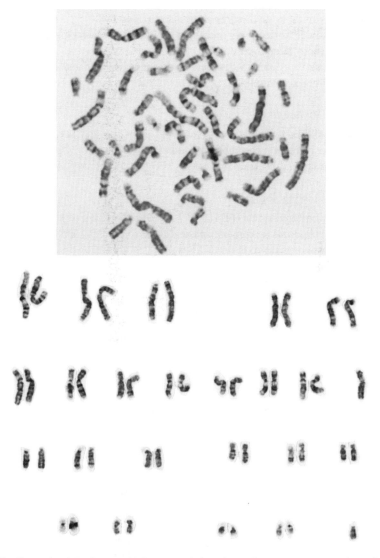

Figure 2b. Giemsa-banded cell (top) with karyotype (below) from a bone marrow preparation stained with Leishman as detailed in Section 4.2.3.

(iii) Trypsin-EDTA 10× concentrate [trypsin (1:250) 5 g/l with EDTA 2 g/l] can be made in the laboratory but will vary from batch to batch as compared with purchased solution. Store in the freezer.

(iv) Working staining solution (made day of use) is prepared by mixing 26 ml buffer (pH 6.8), 7 ml methanol, 0.6 − 1.0 ml 10× trypsin-EDTA and 0.8 ml Giemsa stain.

Procedure.

(i) Incubate the slides for 8 min in buffer (pH 6.8) at 56°C.

(ii)   Rinse the slides in distilled or de-ionised water.
(iii)   Place the slides horizontally and flood them with the working stain solution for 8 min at room temperature.
(iv)   Gently rinse the slides in distilled or de-ionised water and air-dry.
(v)   View without a coverslip.

4.2.2 *Method 2. G-Banding*

This procedure allows for greater chromosome digestion and is consequently more suitable for higher resolution chromosome studies (see Chapter 2). Optimal trypsin treatment time is determined by trial and error. The method is similar to that originally described by Seabright (4).

Prepare the following solutions.

(i)   Phosphate-buffered saline (PBS) (pH 7.0). Dissolve 8.0 g of NaCl, 0.2 g of KCl, 0.92 g of anhydrous sodium phosphate dibasic (Na_2HPO_4) and 0.2 g of anhydrous potassium phosphate monobasic (KH_2PO_4) in 1 litre of distilled water.
(ii)   0.005% Trypsin solution. Dissolve 35 mg of trypsin 1:250 (Difco) in 70 ml of PBS. The solution is stable for approximately one day.
(iii)   Giemsa stain. 2.5 ml of Giemsa (Gurr's) plus 45 ml of phosphate buffer (pH 6.8) (see Section 4.1.2 for details).

Procedure.

(i)   Incubate the slides for 20−40 sec in the trypsin solution in a Coplin jar.
(ii)   Rinse the slides thoroughly with cold (refrigerated) PBS.
(iii)   Stain for 5 min in Giemsa solution.
(iv)   Rinse the slides in distilled water and air-dry.

4.2.3 *Method 3. G-Banding (Leishman G-Banding)*

Some laboratories find that the use of Leishman stain instead of Giemsa aids the G-banding of difficult chromosomes such as those obtained from direct bone marrow preparations. This stain tends to counteract the 'fuzziness' of morphology often inherent in direct preparations.

Prepare the following solutions.

(i)   Leishman stain concentrate. Add 0.3 g of Leishman stain to 200 ml of Analar methanol in a volumetric flask. Place the flask on a hotplate at 70°C overnight. Filter the solution and decant into aliquots as required. Dilute 1:5 stain to buffer (pH 6.8) for use.
(ii)   Trypsin. Reconstitute a vial of Bacto-trypsin (Difco) with 10 ml of de-ionised water. Add 2 ml of this to 40 ml of normal saline. Keep the remainder as a stock solution for up to 5 days.
(iii)   Buffer (pH 6.8). Make as in Section 4.1.2 or use Gurr's buffer tablets (pH 6.8) dissolved in de-ionised water.
(iv)   Normal saline. May be obtained commercially or made up (0.9% w/v).

Procedure.

(i)   Place the slides (preferably at least 2 days old) into a Coplin jar of trypsin solution for 5−20 sec depending on the age of the slides and variations of each batch of trypsin. Slides from lymphocyte cultures will need less time than those from amniotic fluid or fibroblast cultures.

(ii)   Rinse in a Coplin jar of saline.

(iii)   Rinse in a Coplin jar of buffer (pH 6.8).

(iv)   Place the slides horizontally on a rack over a sink and stain with the Leishman stain solution for 5 min.

(v)   Rinse the slides in buffer (pH 6.8).

(vi)   Rinse in distilled water.

(vii)   Dry on a hotplate at 40°C.

4.3 Quinacrine Banding

Quinacrine dihydrochloride (quinacrine, atebrin) is an acridine dye which binds to DNA either by intercalation or by external ionic binding (9). While the base composition of isolated chromatin fractions can affect fluorescence of this dye (10,11), the banding patterns of whole chromosomes appear to be influenced strongly by variation in the protein composition of the chromosome (7).

As a tool for identifying chromosomes, quinacrine banding (Q-banding) is of great value (12). For human chromosomes, the banding patterns strongly resemble those seen in G-banding (*Figure 3*), although notable exceptional regions are the centromeric regions of chromosomes 1, 9 and 16 and the acrocentric satellite regions. Polymorphic variation in these regions and variations of the Y chromosome can often be recognised using this staining.

The procedure for Q-banding is simple, quick and reliable without the need to age or pre-treat slides before staining. The method works equally well with old slides, and even poor preparations will show some banding. The method described is similar to that of Uchida and Lin (13).

4.3.1 *Method for Q-banding*

Prepare the following solutions.

(i)   Quinacrine dihydrochloride solution. Dissolve 0.5 g of quinacrine dihydrochloride in 100 ml of distilled water. Store the solution in a foil-covered container in the refrigerator.

(ii)   MacIlvaine's buffer (pH 5.6). Make a solution of 0.1 M anhydrous citric acid (19.2 g/l) (solution A) and a solution of 0.4 M anhydrous sodium phosphate dibasic (Na_2HPO_4) (56.8 g/l) (solution B). The buffer solution consists of 92 ml of solution A and 50 ml of solution B, pH adjusted to 5.6 if necessary.

Procedure

(i)   Place the slide in quinacrine stain for 10 min.

(ii)   Rinse briefly in tap water to remove excess stain.

(iii)   Place in the buffer for 1−2 min.

(iv)   Mount with a few drops of buffer using an extra thin coverslip (thickness 0, Clay Adams). Remove as much buffer as possible from between the coverslip and the slide by gently squeezing the slide between paper towels or bibulous paper.

(v)   Seal the coverslip to the slide with rubber cement or nail varnish, if desired.

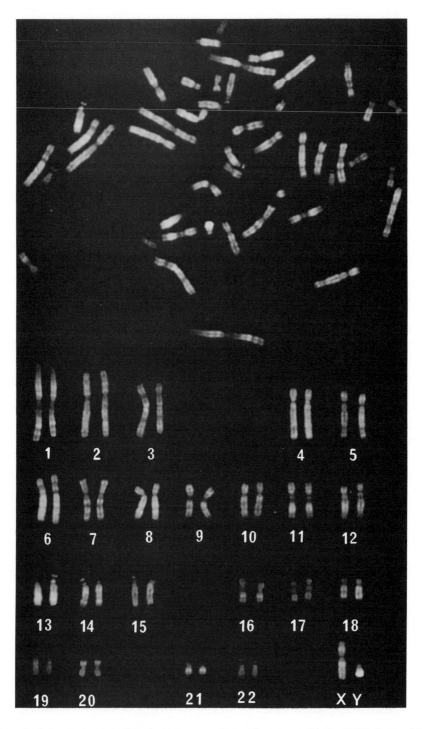

Figure 3. Quinacrine banded cell (top) with karyotype (below) from a normal male, 46,XY (Section 4.3.1).

(vi) Scan immediately with the fluorescent microscope (see Section 2 for appropriate filter combinations).

(vii) Photography (Section 3) should be carried out promptly before images significantly fade.

After performing this staining technique the coverslips can be removed by cutting through the seal and soaking the slide in running tap water for several minutes. Dried slides can be subsequently stained by another technique after de-staining as described in Section 4.1.2. Slides can also be re-stained with quinacrine.

4.4 Constitutive Heterochromatin Banding

Constitutive heterochromatin is the structural chromosomal material seen as dark staining material in interphase as well as during mitosis. It includes both repetitive DNA, satellite DNA (as detected on centrifugation gradients) and some non-repetitive DNA. C-banding is said to demonstrate constitutive heterochromatin (*Figure 4*) since satellite DNA has been localised by *in situ* hybridisation to darkly staining C-band regions (14).

For human chromosomes, darkly staining C-bands are located at the centromeres of the chromosomes, with the exception of the Y chromosome where the dark band is located on the distal region of the long arm. Marked polymorphism is present in the size of C-bands. Thus, C-banding has application for investigating chromosome rearrangement near centromeres and in investigating polymorphism.

Procedures used in *in situ* hybridisation involve treatment with acid, alkali and hot saline citrate (14). When preparations are subsequently stained with Giemsa, C-bands are visualised, leading to the explanation that DNA denaturation and re-annealing was responsible for C-bands. However, these treatments also result in the removal of DNA from the chromosomes and this appears to account for the differential staining in C-banding (15).

4.4.1 *Method for C-Banding*

This method uses a saturated solution of barium hydroxide as the denaturing agent for C-banding and is a modification of that described by Salamanda and Armendares (16).
Prepare the following solutions.

(i) Barium hydroxide [$Ba(OH)_2$] solution. Dissolve 11.04 g/l and store in an airtight bottle until immediately prior to use.

(ii) 2 × SSC solution. Mix 1 part of 0.03 M sodium citrate (17.4 g/l) with 1 part of 0.03 M NaCl (8.82 g/l).

(iii) Make up Giemsa solution as in Section 4.1.2.

(iv) 0.2 M HCl.

Procedure.

(i) Pre-heat a Coplin jar with filtered $Ba(OH)_2$ solution to 37°C and a Coplin jar of 2 × SSC to 65°C.

(ii) Treat the slides in 0.2 M HCl at room temperature for 30 min.

(iii) Rinse the slides twice in distilled or de-ionised water.

(iv) Treat the slides with 0.07 M $Ba(OH)_2$ for 10 min at 37°C.

(v) Rinse the slides in distilled or de-ionised water three times.

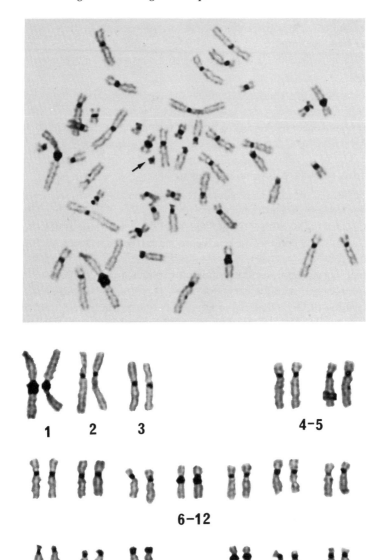

Figure 4. C-banded cell (top) with karyotype (below) from a phenotypically normal female. A small additional marker chromosome (arrow) is present (46,X + mar). The marker is seen to be composed of both darkly staining and lighter staining material.

(vi)   Incubate the slides in 2 × SSC at 65°C for 2 h.
(vii)   Rinse the slides in distilled or de-ionised water.
(viii) Stain the slides in Giemsa solution for 1−2 h.
(ix)   Rinse the slides and air-dry them.

4.5 Reverse Banding

Bands that are negative, appearing pale by G-banding, stain darkly by R-banding. Conversely, dark positive G-bands appear pale using R-banding techniques (*Figure 5*). R-banding can be achieved by incubation in hot saline solution followed by Giemsa staining (17). A second method involves staining with acridine orange (18), the resulting bands being green and red in colour. As previously indicated for G-bands, the pattern seen reflects the structural and functional composition of chromosomes (6). However, the chemical basis for the staining reactions remains obscure.

One elegant method described below involves staining with the fluorescent dye chromomycin A_3 followed by counterstaining with a second dye, methyl green, that has a complementary base pair binding specificity (19). Energy transfer is thought to suppress fluorescence except in regions where DNA base pairs are predominantly of one type (19).

4.5.1 *Method 1. R-Banding Using Chromomycin A_3*

This method can be successfully carried out on old slides as well as fresh preparations.
    Prepare the following solutions.

(i)   0.14 M phosphate buffer (pH 6.8) containing 500 μM magnesium chloride.
        Dissolve 19.3 g of sodium phosphate monobasic monohydrate ($NaH_2PO_4H_2O$)
        in 1 litre of distilled water (solution A). Dissolve 19.9 g of anhydrous sodium
        phosphate dibasic (Na_2HPO_4) in 1 litre of distilled water (solution B). Titrate
        solution B with solution A to pH 6.8. Add 10 mg of $MgCl_2H_2O$ for every 100 ml
        of pH 6.8 buffer.
(ii)   100 μM chromomycin A_3 in phosphate buffer (pH 6.8). Dissolve 1 mg of
        chromomycin A_3 in 1 ml of the buffer described above. Store this stock solution
        in the freezer. Prepare a working solution by diluting 0.3 ml of stock with 0.2 ml
        of phosphate buffer.
(iii)   0.15 M NaCl in 0.005 M Hepes buffer (pH 7.0). To 1.75 g of NaCl and 238
        mg of Hepes add distilled water to give a final volume of 200 ml. Adjust to pH
        7.0 with 1 M NaOH.
(iv)   100 μM methyl green in 0.15 M NaCl/0.005 M Hepes. Dissolve 2.6 mg of methyl
        green in 50 ml of NaCl/Hepes buffer. This solution should be prepared freshly
        the day of use since methyl green is unstable in solution.

Procedure.

(i)   Layer the slide with 0.14 M phosphate buffer (pH 6.8) containing 500 μM MgCl
        for 10 min at room temperature.
(ii)   Discard the buffer and place 3−4 drops of chromomycin A_3 solution on the slide.
(iii)   Coverslip and stain for 10 min at room temperature.
(iv)   Rinse the slide with 0.15 M NaCl/0.005 M Hepes buffer.

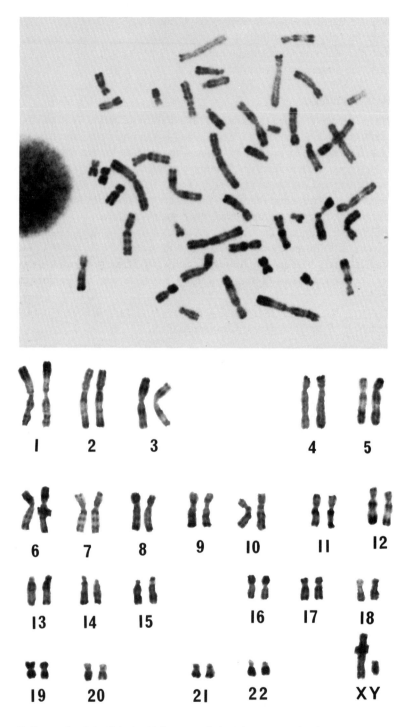

Figure 5. Reverse-banded cell (top) with karyotype (below) from a normal male, 46,XY, (Section 4.5.2).

(v) Place 4 – 6 drops of 100 μM methyl green on the slide, coverslip and stain for 10 min at room temperature (or place the slide in a Coplin jar of methyl green).
(vi) Rinse the slide with NaCl/0.005 M Hepes buffer.
(vii) Mount a coverslip on the slide using glycerol as a mounting medium.
(viii) View under the fluorescence microscope as described in Section 2.

Note that glycerol retards the fading of the chromomycin A_3 fluorescence. Better fluorescence is sometimes obtained one day after staining. Slides can be de-stained as described previously (Section 4.1.2) and re-stained by a second method.

4.5.2 *Method 2. R-Banding with Acridine Orange*

This method is most successful with slides aged at room temperature for 1 – 2 weeks. The protocol is based on that described by Verma and Lubs (18).
 Prepare the following solutions.
(i) Phosphate buffer (pH 6.5). Dissolve 9.93 g of anhydrous sodium phosphate dibasic (Na_2HPO_4) in 1 litre of distilled water (solution A) and 9.1 g of potassium phosphate dibasic (KH_2PO_4) in 1 litre of distilled water (solution B). To make the buffer mix 32 ml of solution A with 68 ml of solution B. Adjust to pH 6.5 with additional solution A if necessary.
(ii) 0.01 % Acridine orange in phosphate buffer. Dissolve 10 mg of acridine orange in 100 ml of phosphate buffer (pH 6.5). Store in a darkened container in the refrigerator.

Procedure.
(i) Pre-heat a Coplin jar containing phosphate buffer (pH 6.5) to 85°C.
(ii) Incubate the slide for 8 – 10 min in the buffer at 85°C. Shorter incubation time is usually required for older slides.
(iii) Stain the slide in 0.01 % acridine orange at room temperature for 5 min.
(iv) Rinse the slide in buffer (pH 6.5).
(v) Mount the slide in buffer (pH 6.5) [as in Section 4.3.1(v)] and view with the fluorescent microscope.
(vi) Photography can be carried out as described previously. However, best results are obtained using Kodachrome II colour transparency film with an ASA setting of 320 (DIN 26). Processed transparencies can be printed on black and white paper.

4.6 Nucleolar Organiser Region Staining

The nucleolar organiser regions of mammalian chromosomes are known to contain the genes for 18S and 28S rRNA (20). Regions in which the genes are thought to be actively transcribed can be selectively stained using silver nitrate (NOR-staining) (*Figure 6*). The stain is thought to selectively identify a protein adjacent to the nucleolar organiser region rather than the nucleolar organiser regions themselves (21).
 For human chromosomes the genes for 18S and 28S rRNA are located on the short arms of the acrocentrics (chromosomes 13, 14, 15, 21 and 22). Silver staining will selectively identify these regions although not all acrocentrics will stain and chromosomes from different individuals will stain more or less intensely in a consistent manner. Thus,

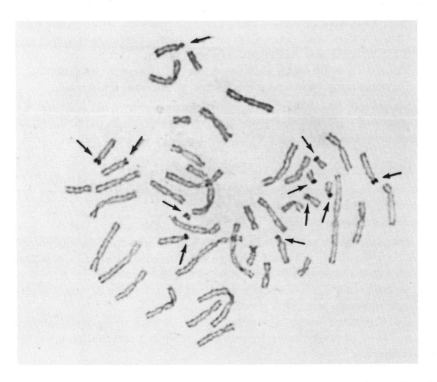

Figure 6. A cell stained by NOR-banding. Acrocentic chromosomes, some of which show positive, dark, staining at the nucleolar organiser regions, are marked by arrows. Also marked by an arrow is a small marker chromosome (as in *Figure 4*) which stains weakly with silver staining.

a specific staining pattern will be seen for each individual and this polymorphism is a heritable characteristic. Silver staining therefore has a variety of applications for research and clinical genetics (22).

4.6.1 *Method 1. NOR-Banding*

This method is very simple although somewhat unreliable (23). Some batches of silver nitrate appear to work better than others and some preparations stain better than others. The method does have the substantial advantage of allowing an additional banding technique to be carried out after the silver staining has been completed. Thus, silver staining, Q-banding, R-banding and G-banding could all be carried out on the same cell.

The only solution required is a 50% silver nitrate solution (5.0 g $AgNO_3$ in 10 ml of distilled water). Solution should be stored in a dark bottle in the refrigerator and should not be used if a black precipitate is present.

Procedure.

(i)   Place wet filter paper or a paper towel in the bottom of a Petri dish or an appropriate container.

(ii) Place the slide on top of the wet paper and carefully cover with silver nitrate solution. Place a coverslip on the slide. Incubate for up to 24 h at 37°C. Avoid exposure to the light.

(iii)   Rinse the slide well with water.

(iv)   Stain the slide with Giemsa or another stain of choice.

4.6.2 *Method 2. NOR-Banding*

This method is based on that of Goodpasture and Bloom (24).
  Prepare the following solutions.

(i)   A 50% silver nitrate solution as described in Section 4.6.1.

(ii)   3% formalin (pH 4.5). Add 3 ml of formalin (or 1.0 ml of 0.37% formaldehyde) to 97 ml of distilled water (or 99 ml of distilled water if formaldehyde is used). Adjust the pH to 7.0 with 1 M sodium acetate. Then adjust the pH to 4.5 with concentrated formic acid. Store the solution in the refrigerator.

(iii)   Ammoniacal silver. Dissolve 4.0 g of silver nitrate in 5 ml of concentrated ammonium hydroxide. Slowly add 7.5 ml of distilled water. There should be no precipitate. Store in a dark bottle in the refrigerator.

Procedure.

(i)   Place three drops of 50% silver nitrate on the slide. Coverslip and place in an oven at 60°C until the silver nitrate is crystalline.

(ii)   Rinse the slide in distilled water to remove the coverslip.

(iii)   Place three drops of 3% formalin solution and 1–3 drops of ammoniacal silver solution on the slide. Coverslip the slide.

(iv)   Observe the staining reaction under the microscope at low power (10×). When the cells have developed a golden brown colour (30–60 sec, typically), rinse the slide well in distilled water.

(v)   Air-dry the slides and observe at high power.

4.7 **Differential Replication Staining**

From studies incorporating [³H]thymidine followed by autoradiography it has been known for many years that different parts of mammalian chromosomes replicate at different times (25). With the introduction of methods which detect the incorporation of 5-bromodeoxyuridine (BrdU) into chromosomes, resolution of differential replication has been considerably enhanced (26).

  BrdU is a thymidine analogue which is readily incorporated into chromosomes. When such BrdU-substituted chromosomes are subsequently stained with the bisbenzimidazole dye Hoechst 33258 (a compound which binds to DNA at A+T base pairs) there is a quenching of the bright fluorescence normally seen when unsubstituted chromosomes are stained with this dye. The quenching effect can be visualised by fluorescence microscopy of the fixed chromosome preparations (27) or the preparations can be subsequently exposed to photolysis and stained with Giemsa (FPG technique) (28).

  The approach can be applied to the detection of different cycles of replication or to differentiate between early and late replications within a single cell cycle. When BrdU is added for the last part of one cell cycle (B-pulse), chromosome regions replicating early will stain brightly (darkly with Giemsa) and the resulting pattern closely resembles R-bands (*Figure 7a*). Conversely, when BrdU is present early in one cell cycle (T-pulse), chromosome regions replicating late will stain brightly (darkly with Giemsa)

73

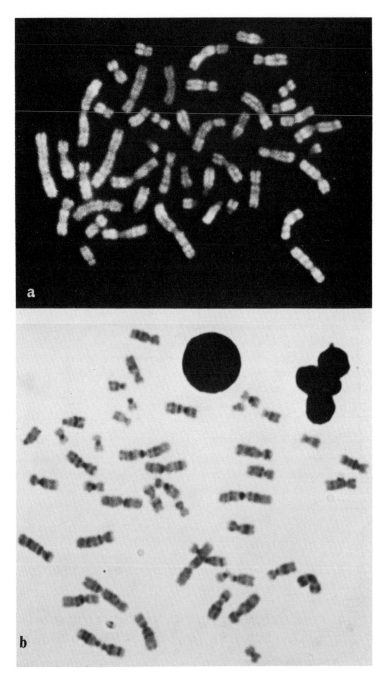

Figure 7. Differential replication staining. **(a)** BrdU present for the last part of one cell cycle (B-pulse) resulting in bands similar to R-bands. The cell is stained with Hoechst 33258 and viewed directly under the fluorescence microscope (Section 4.7.1). **(b)** BrdU present for the early part of one cell cycle (T-pulse) giving a G-band appearance to the chromosomes. The cell is stained with Hoechst 33258, exposed to photolysis and Giemsa stained (Section 4.7.2).

$$X_e \quad X_l$$

BUdR EARLY

BUdR LATE

Figure 8. X-chromosome from normal female cells labelled with BrdU by B-pulse and T-pulse protocols. Late replicating X-chromosomes are readily distinguishable from early replicating X-chromosomes.

and the pattern resembles G-bands *(Figure 7b)*. Homologues usually have similar banding patterns, a notable exception being the X chromosome *(Figure 8)*. Cells which have undergone one cell cycle in the presence of BrdU may show differential staining between chromatids at certain bands (26). This lateral asymmetry reflects regions in which one chromatid contains much more thymine than does the other chromatid. Differential replication staining can be combined with higher resolution analysis for detailed analysis of chromosome replication (29). When BrdU is present for two full cycles, one chromatid contains DNA with BrdU substituted into one polynucleotide chain while the sister chromatid will contain BrdU substituted into both chains. The resulting difference in staining is the basis of the detection of sister chromatid exchanges *(Figure 9)*.

Three protocols are provided for the observation of different times of replication. The procedures are designed for peripheral blood cultures (see Chapter 2). Some adjustment in times are necessary for studies on other cell types. When cell culture contains BrdU, deoxycytidine is also added to reduce toxicity of the BrdU (30). Fluorodeoxyuridine (FUdR) at a final concnetration of 0.4 μM and uridine at a final concentration of 6 μM are also frequently added at the same time as BrdU. FUdR inhibits thymidylate synthetase and thus reduces *de novo* synthesis of deoxythymidine-5'-monophosphate leading to greater BrdU utilisation. Satisfactory incorporation of BrdU can however be obtained without the addition of FUdR and uridine. Cultures containing BrdU should be protected from the light (for example, by wrapping in aluminium foil).

4.7.1 *Replication Bands − B Pulse*

Prepare the following solutions.

(i) 10^{-2} M stock BrdU. Dissolve 30.7 mg of BrdU in 10 ml of distilled water. Sterilise by micropore filtration and store frozen.

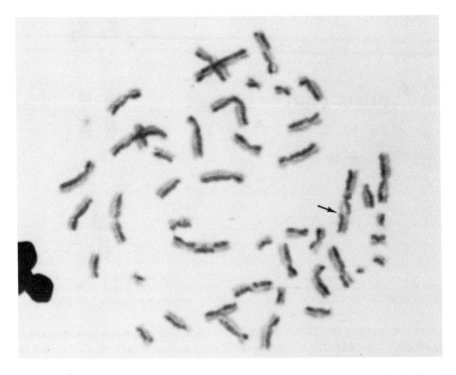

Figure 9. Differential staining with BrdU present for two cell cycles to allow identification of each chromatid (Section 4.7.3). Staining is as described in *Figure 7b*. Sister chromatid exchange is illustrated by the arrow.

(ii) 10^{-2} M stock deoxycytidine (dC). Dissolve 28.0 mg of deoxycytidine hydrochloride in 10 ml of distilled water. Sterilise by micropore filtration and store frozen.

(iii)   Cell culture media (e.g., Eagle's minimal essential medium, MEM) with 20% fetal bovine serum and phytohaemagglutinin (PHA) as described in Chapter 2.

(iv)   Chromosome harvest reagents, hypotonic and fixative as described in Chapter 2.

(v)   Stock 50 μg/ml Hoechst 33258 in distilled water. Working strength 0.5 μg/ml Hoechst 33258 is made by diluting one part of stock in 100 parts of PBS (see Section 4.2.2 for formulation of PBS).

(vi)   MacIlvaine's buffer (pH 7.5). Make a 0.1 M anhydrous citric acid solution (19.2 g/l) (solution A) and 0.2 M anhydrous sodium phosphate dibasic (Na_2HPO_4) solution (28.4 g/l) (solution B). Buffer at pH 7.5 is obtained by mixing 80 ml of A and 920 ml of B and adjusting the pH to 7.5 as necessary.

(vii)   2 × SSC solution. See Section 4.4.1 for details.

(viii)   5−10% Giemsa in pH 6.8 phosphate buffer. See Section 4.1.2 for details.

Procedure.

(i)   Initiate 10 ml peripheral blood cultures as described in Chapter 2.

(ii)   5−7 h before the addition of colcemid, add 0.1 ml of BrdU stock (final concentration 10^{-4} M) and 0.1 ml of dC stock (final concentration 10^{-4} M).

(iii)   One hour before harvest add 0.1 ml of colcemid stock at a concentration of 10 μg/ml.

(iv)   Harvest and make slides as described in Chapter 2.

(v)   Stain the slides. Two methods are widely used. Firstly, soak the slides in PBS for 5 min; stain in 0.5 μg/ml Hoechst 33258 for 10 min; rinse briefly in PBS; rinse briefly in distilled water; mount the slides with MacIlvaine's buffer (pH 7.5) and view under the fluorescence microscope as described in Section 2. Alternatively, stain the slides with Hoechst 33258 and mount in McIlvaine's buffer as immediately described above; irradiate the slides approximately 5 cm from a 15 W blacklight source at 50°C for 15 min or a 75 W growlamp for 24 h; remove the coverslips; rinse the slides in distilled water; incubate the slides for 15 min in 2 × SSC at 65°C; stain the slides for 5−10 min in Giemsa.

4.7.2 *Replication Bands − T Pulse*

All the solutions listed in Section 4.7.1 are required. In addition, prepare 10^{-3} M stock thymidine (T). Dissolve 2.42 mg of thymidine in 10 ml of distilled water. Sterilise by micropore filtration and store frozen.

Procedure.

(i)   Initiate 10 ml peripheral blood cultures. Add 0.1 ml of BrdU stock and 0.1 ml of dC stock at the beginning of cell culture.

(ii)   After 42 h, remove the BrdU/dC containing media, wash the cells once in tissue cutlure media with serum, and add fresh media. Also add 0.1 ml of T stock to the 10 ml culture to give a final concentration of 10^{-5} M T.

(iii)   After a further 5−6 h add colcemid as described in Section 4.7.1.

(iv)   Continue as in Section 4.7.1.

4.7.3 *Replication Staining − Sister Chromatid Exchange*

Solutions as listed in Section 4.7.1 are required.

Procedure.

(i)   Initiate 10 ml peripheral blood cultures. Add 0.1 ml of BrdU stock and 0.1 ml of dC stock at the beginning of cell culture.

(ii)   Maintain the cultures for 48−72 h.

(iii)   Add colcemid, harvest and stain as in Section 4.7.1.

4.8 **DAPI/Distamycin A Staining**

DAPI (4,6-diamino-2-phenyl-indole) is a fluorescent dye with DNA affinity for A+T-specific binding. When chromosomes are stained with this compound a banding pattern similar to Q-banding is seen although contrast is poor. Distamycin A also has an affinity for A+T-specific DNA but is non-fluorescent. The two dyes have similar base pair binding preference but non-identical binding affinities and dissimilar structures. When chromosomes are stained with DAPI (the primary stain) followed by counterstaining with distamycin A, a specific subset of chromosome bands are highlighted. The heterochromatic regions of chromosomes 1, 9, 16, the distal long arm of the Y chromosome and the proximal short arm of chromosome 15 stain brightly while all

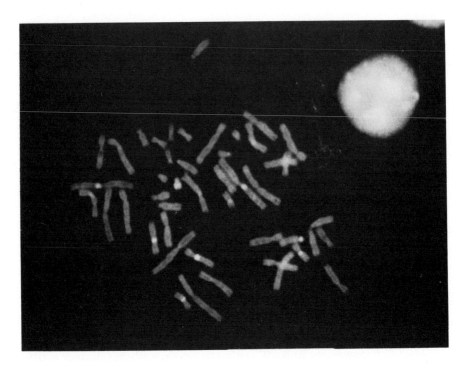

Figure 10. DAPI/distamycin A staining using the method given in Section 4.8.1. Bright fluorescence is seen at bands of the centromeres of chromosomes 1, 9, 16, the short arm of chromosome 15 and the distal long arm of the Y chromosome.

other regions appear dull (31). This particular staining technique (DAPI/DA) can also be obtained with other compounds with similar binding specificities. For example, Hoechst 33258 can be substituted for DAPI and methyl green can be used in place of distamycin-A (32). The use of fluorescent and non-fluorescent dye combinations has been reviewed by Schweizer (33).

DAPI/DA staining can be useful in identifying the chromosome regions mentioned above (*Figure 10*), notably abnormalities and variants of chromosome 15. Because of rapid fading of the fluorescence, photographing DAPI-DA can be problematical.

4.8.1 *Method 1 for DAPI/DA Bands*

Prepare the following solutions.

(i)   MacIlvaine's buffer (pH 7.5). See Section 4.7.1 for the method to make this buffer.

(ii)   A stock solution of distamycin A is made by dissolving 2 mg in 10 ml of McIlvaine's buffer (pH 7.5). The stock solution should be stored in a freezer. Working solution is a 1 in 10 dilution of the stock in McIlvaine's buffer (i.e., final concentration 10 μg/ml).

(iii)   A stock solution of DAPI is made by dissolving 2 mg in 10 ml of distilled water (a small amount of methanol aids in dissolving DAPI). Store the stock solution in a freezer. Working strength solution (0.2 μg/ml) is a 1 in 1000 dilution of the stock in MacIlvaine's buffer.

Procedure.

(i)   Soak the slide in MacIlvaine's buffer for 10 min.
(ii)   Stain the slide for 10 min with distamycin A (20 μg/ml).
(iii)   Rinse the slide briefly in MacIlvaine's buffer.
(iv)   Stain the slide for 10 min in DAPI (0.2 μg/ml).
(v)   Rinse the slide briefly in MacIlvaine's buffer.
(vi)   Mount the slide with a 1:1 dilution of glycerol with MacIlvaine's buffer.
(vii)   View under the fluorescence microscope with filters as described in Section 2.

If the chromosomes show a Q-banding-like appearance, stain longer with distamycin A. If the chromosomes appear to be too dull, try reducing the distamycin A staining time.

4.8.2 *Method 2 for DAPI-DA Bands (Hoechst/DA Bands)*

Prepare the following solutions.

(i)   PBS (pH 7.0). See Section 4.2.2 for formulation.
(ii)   0.5 μg/ml Hoechst 33258 solution in PBS. See Section 4.7.1.
(iii)   MacIlvaine's buffer as in Section 4.8.1.
(iv)   Distamycin A solution as in Section 4.8.1.

Procedure.

(i)   Soak the slide in PBS for 5 min.
(ii)   Stain the slide for 10 min in 0.5 μg/ml Hoechst 33258.
(iii)   Rinse the slide for 6 min in PBS.
(iv)   Rinse the slide in water and air-dry.
(v)   Place three drops of distamycin A 20 μg/ml on the slide, coverslip and stain for $2-5$ min at room temperature (optimal time tends to be somewhat variable).
(vi)   Rinse the slide in MacIlvaine's buffer (pH 7.5) and air-dry.
(vii)   Mount with a coverslip and view as described in Section 4.8.1.

4.9 Other Banding Techniques

A number of other banding techniques exist. Many of these are of research interest rather than of routine clinical use. They are, however, briefly reviewed here for completeness.

4.9.1 *Telomere Banding*

The most distal regions of the chromosome arms, the terminal bands or telomeric regions, can be selectively stained by T-banding (34). The staining can be considered to be a special case of R-banding in which a more destructive treatment results in diminished staining except at terminal bands.

Methods to obtain T-bands are described by Dutrillaux (35). One method involves heating 94 ml of distilled water and 3 ml of phosphate buffer pH 6.7 to 87°C. A few minutes before staining, 3 ml of Giemsa are added and the slides are then stained for

5 – 30 min. Slides can be destained by passing them through alcohol/water washes and then re-stained with acridine orange [5.0 mg in 100 ml phosphate buffer (pH 6.7)] with analysis carried out by fluorescence microscopy.

A second method described by Dutrillaux (35) requires a 20 – 60 min incubation at 87°C in Earle's balanced salt solution, PBS or phosphate buffer (pH 5.1).

4.9.2 *G-11 Banding*

G-11 banding simply involves staining chromosome preparations with Giemsa at pH 11.0 (36). Human chromosomes stain blue with the centromeric regions of chromosomes (notably chromosome 9) staining purplish-red (magenta). Rodent chromosomes stain uniformally magenta. The application of this technique is the identification of chromosomes in man-mouse somatic cell hybrids (37) (see Chapter 7). The mechanism of G-11 staining is uncertain (38).

Giemsa is diluted one part plus 49 parts distilled water and the pH adjusted to 11.0 with any suitable alkali solution. The stain is warmed to 37°C and slides are stained for 10 – 20 min.

An alternative method is described in detail in Chapter 7, Section 3.8.1.

4.9.3 *Kinetochore Staining*

The kinetochore is the point of attachment of a chromosome to the mitotic spindle, located at the centromeric constriction of the chromosome. A number of techniques exist to identify pairs of dots at the centromeres of chromosomes (Cd-banding) (7). These dots may represent the kinetochore itself or chromatin associated with this structure.

Cd-banding is obtained by a method described by Eiberg (39). Cell cultures and harvest are as described in Chapter 1 except that the first fixative contains methanol and acetic acid in the ratio 9:1. A second fixation reduces the proportions of methanol and acetic acid to 5:1 and a third fixation consists of 3:1 methanol and acetic acid. Slides are aged for 1 week at room temperature. Banding is induced by incubating the slides in Earle's balanced salt solution at pH 8.5 – 9.0 at 85°C for 45 min. Slides are stained in 4% Giemsa in phosphate buffer (pH 6.5) for 10 min.

An alternative method stains both nucleolar organiser regions and centromeric dots simultaneously (40). Standard preparations are treated for 30 – 40 sec with 0.01% aqueous NaOH solution (pH 8.5), (approximately 10^{-5} M NaOH), rinsed, and stained by the ammoniacal silver technique (Section 4.6.2).

5. SEX CHROMATIN IDENTIFICATION

In clinical genetics it is sometimes necessary to attempt to rapidly determine the sex of an individual without waiting for a peripheral blood chromosome analysis. Methods to identify sex chromatin tend to be somewhat unreliable – artefacts can result in misleading results, some chromosome abnormalities and mosaics being missed. It is therefore always necessary to confirm any preliminary results with a cytogenetic analysis.

X-chromatin can be identified in interphase cells stained with orcein as a dark body (chromocentre) or 'Barr body' (41) (*Figure 11a*). The number of Barr bodies in any

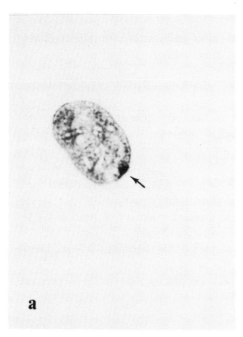

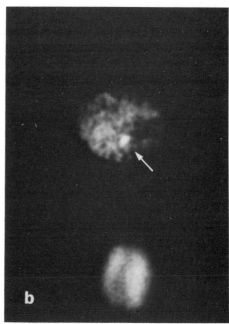

Figure 11. Sex chromosome identification. **(a)** Typical Barr body (Section 5.1). **(b)** Fluorescent Y-body (Section 5.2).

given cell relates to the number of X chromosomes present. In theory, the number of such bodies is equal to the number of inactivated X chromosomes, i.e., the total number of X chromosomes, minus one. Thus in a normal male cell, no Barr bodies are seen, in a normal female cell one Barr body is present and in a XXX female two Barr bodies are present. In practice, the number of Barr body-positive cells for a normal female will be in the range of $50 - 80\%$, somewhat less $(30 - 60\%)$ for preparations from buccal mucosa and even lower for newborn normal females (42). A low percentage (up to 5%) of apparently positive cells may also be seen in preparations from normal males.

Y-chromatin can be seen in the interphase cells of males stained with quinacrine dihydrochloride (43). A brightly fluorescent spot in an interphase cell corresponds to the bright band on the distal long arm of the Y chromosome in metaphase preparations. Note that other brightly staining Q-bands (some of which may show marked polymorphism) can lead to misinterpretation of results. Some Y chromosomes show a relatively small brightly fluorescent band and this may be difficult to recognise in an interphase cell.

Sex chromatin identification is usually carried out on smears of buccal mucosa. To prepare the smear, clean the inside of the cheek with a piece of gauze. Scrape the cheek with the edge of a wood tongue depressor or metal spatula until some white cellular material can be seen on the blade. Spread onto a slide in one smooth smear. Usually, a preparation is made from each cheek.

Stain the preparations immediately for X- or Y-chromatin bodies as detailed below:

5.1 X-Chromatin Bodies

(i)   Place 3−4 drops of freshly filtered 2% aceto-orcein stain on the slide (see Section 4.1.1 for detailed information on aceto-orcein).

(ii)   Cover with a coverslip.

(iii)   Squeeze out excess stain by pressing the slide between paper towel or bibulous paper.

(iv)   Examine immediately under the light microscope. Score at least 100 cells for the presence or absence of a Barr body.

5.2 Y-Chromatin Bodies

(i)   Soak the freshly prepared slide in a Coplin jar with absolute methanol for a minimum of 20 min (or several hours if necessary).

(ii)   Stain for 5−7 min in 0.5% quinacrine mustard solution (see Section 4.3.1).

(iii)   Rinse the slide briefly in water to remove excess stain.

(iv)   Rinse the slide briefly in MacIlvaine's buffer (pH 5.6) (Section 4.3.1) and mount the slide in buffer.

(v)   View under the fluorescence microscope. Score at least 100 cells for the presence or absence of a Y-body.

6. NOMENCLATURE

Codes to describe banding techniques have been widely used to summarise the methodology applied to the identification of a particular chromosome or polymorphism. A three letter code (triplet) consists of a first letter to denote the type of banding, the second letter the general technique and the third letter denotes the stain (44). For example:

```
Q-   = Q-bands
QF-  = Q-bands by fluorescence
QFQ  = Q-bands by fluorescence using quinacrine
QFH  = Q-bands by fluorescence using Hoechst 33258

G-   = G-bands
GT-  = G-bands by trypsin
GTG  = G-bands by trypsin using Giemsa
GTL  = G-bands by trypsin using Leishman
GAG  = G-bands by acetic saline using Giemsa

C-   = C-bands
CB-  = C-bands by barium hydroxide
CBG  = C-bands by barium hydroxide using Giemsa

R-   = R-bands
RF-  = R-bands by fluorescence
RFA  = R-bands by fluorescence using acridine orange
RH-  = R-bands by heating
RHG  = R-bands by heating using Giemsa
RB-  = R-bands by BrdU
RBG  = R-bands by BrdU using Giemsa
```

RBA = R-bands by BrdU using acridine orange

T-   = T-bands
TH- = T-bands by heating
THG = T-bands by heating with Giemsa
THA = T-bands by heating with acridine orange

The codes may be used in conjunction with additional abbreviations to denote polymorphic variation (see Chapter 4).

Methods which involve combinations of stains or techniques become difficult to describe using the three letter code. For example, DAPI/distamycin A staining (Section 4.8) involves the use of two stains. For fluorescent dyes used in conjunction with counterstains, Schweitzer (33) has proposed that the 'primary' fluorescent stain is referred to first with the 'secondary' counterstain placed after a single slash (i.e., DAPI/DA). This terminology is suggested irrespective of the order in which the stains are applied to the chromosomes.

7. CHROMOSOME STAINING – AN OVERVIEW

From this chapter it is clear that the cytogeneticist is equipped with an array of staining techniques that greatly facilitate in the identification of chromosomes. The introduction of chromosome banding in the late 1960s and early 1970s proved to be a major component in the further classification of genetic disease, evaluation of mutagenicity and in fundamental genetic research. While it still often remains difficult to identify extra chromosomal material in unbalanced karyotypes, it is to be hoped that new approaches to chromosome identification will be developed. The combination of molecular genetic approaches with advanced cytogenetic techniques is likely to lead to still greater sophistication in clinical genetics.

8. ACKNOWLEDGEMENTS

We thank Dr. Samuel A. Latt (Children's Hospital Medical Center, Boston, MA) for the method described in Section 4.8.2. We also thank Leesa Maccarino for typing and Mitchell Silverstein for assistance in preparing illustrations. This work was supported by Lifecodes Corporation.

9. REFERENCES

1. Barer,R. (1953) *Lecture Notes on the Use of the Microscope*, published by Blackwell Scientific Publications, Oxford.
2. Langford,M. (1984) *The Darkroom Handbook*, published by Alfred A. Knopf, New York.
3. Sumner,A.T. and Evans,H.J. (1973) *Exp. Cell Res.*, **81**, 223.
4. Seabright,M. (1972) *Chromosoma*, **36**, 204.
5. Sumner,A.T., Evans,H.J. and Buckland,K.A. (1971) *Nature New Biol.*, **232**, 31.
6. Holmquist,G., Gray,M., Porter,T. and Jordan,J. (1982) *Cell*, **31**, 121.
7. Sumner,A.T. (1982) *Cancer Genet. Cytogenet.*, **6**, 59.
8. Sun,N.C., Chu,E.H.Y. and Chang,C.C. (1973) *Chromosome Newsletter,* **14**, 26.
9. Blake,A. and Peacocke,A.R. (1968) *Biopolymers*, **6**, 1225.
10. Weisblum,B. and deHaseth,P.L. (1972) *Proc. Natl. Acad. Sci. USA*, **69**, 629.
11. Pachmann,U. and Rigler,R. (1972) *Exp. Cell Res.*, **72**, 602.
12. Caspersson,T., Farber,S., Foley,G.E., Kudynowski,J., Modest,E.J., Simonsson,E., Wagh,U. and Zech,L. (1968) *Exp. Cell Res.*, **49**, 219.

13. Uchida,I.A. and Lin,C.C. (1974) in *Human Chromosome Methodology*, Yunis,J.J. (ed.), Academic Press, New York, p. 47.
14. Pardue,M.L. and Gall,J.G. (1970) *Science (Wash.)*, **168**, 1356.
15. Holmquist,G. (1979) *Chromosoma (Berl)*, **72**, 203.
16. Salamanca,F. and Armendares,S. (1974) *Ann. Genet.*, **17**, 135.
17. Dutrillaux,B. and Lejeune,J. (1971) *C.R. Acad. Sci. Paris. Ser. D.*, **272**, 2638.
18. Verma,R.S. and Lubs,H.A. (1975) *Am. J. Hum. Genet.*, **27**, 110.
19. Sahar,E. and Latt,S.A. (1978) *Proc. Natl. Acad. Sci. USA*, **75**, 5650.
20. Ghosh,S. (1976) *Int. Rev. Cytol.*, **44**, 1.
21. Schwarzacher,H.G., Mikelsaar,A.-V. and Schnedl,W. (1978) *Cytogenet. Cell Genet.*, **20**, 24.
22. Marcovic,V.D., Worton,R.G. and Berg,J.M. (1978) *Hum. Genet.*, **41**, 181.
23. Bloom,S.E. and Goodpasture,C. (1976) *Hum. Genet.*, **34**, 199.
24. Goodpasture,C. and Bloom,S.E. (1975) *Chromosoma*, **53**, 37.
25. Taylor,J.H. (1958) *Genetics*, **43**, 515.
26. Latt,S.A. (1976) *Annu. Rev. Biophys. Bioeng.*, **5**, 1.
27. Latt,S.A. (1973) *Proc. Natl. Acad. Sci. USA*, **70**, 3395.
28. Perry,P. and Wolff,S. (1974) *Nature*, **251**, 156.
29. Meer,B., Hameister,H. and Cerillo,M. (1981) *Chromosoma*, **82**, 315.
30. Meuth,M. and Green,H. (1974) *Cell*, **2**, 109.
31. Schweizer,D., Ambros,P. and Andrle,M. (1978) *Exp. Cell Res.*, **111**, 327.
32. Donlon,T.A. and Magenis,R.E. (1983) *Hum. Genet.*, **65**, 144.
33. Schweizer,D. (1981) *Hum. Genet.*, **57**, 1.
34. Dutrillaux,B. and Covic,M. (1974) *Exp. Cell Res.*, **85**, 143.
35. Dutrillaux,B. (1973) *Chromosoma*, **41**, 395.
36. Bobrow,M., Madan,K. and Pearson,P.L. (1972) *Nature New Biol.*, **238**, 122.
37. Bobrow,M. and Cross,J. (1974) *Nature*, **251**, 77.
38. Wyandt,H.E., Wysham,D.G., Minden,S.K., Anderson,R.S. and Hecht,F. (1976) *Exp. Cell Res.*, **102**, 85.
39. Eiberg,H. (1974) *Nature*, **248**, 55.
40. Denton,T.E., Brooke,W.R. and Howell,W.M. (1977) *Stain Tech.*, **52**, 311.
41. Barr,M.L., Bertram,L.F. and Lindsay,H.A. (1950) *Anat. Record*, **107**, 283.
42. Hamerton,J.L. (1971) in *Human Cytogenetics*, Vol. **1**, Academic Press, NY, p. 131.
43. Pearson,P.L., Bobrow,M. and Vosa,G.C. (1970) *Nature*, **226**, 78.
44. Paris Conference (1971) Supplement (1975) *Standardization in Human Cytogenetics*, Birth Defects Original Article Series XI, **9**, 1975.

CHAPTER 4

Analysis and Interpretation of Human Chromosome Preparations

J.A.JONASSON

1. INTRODUCTION

In this chapter only a short, oversimplified and perhaps, at times, a too personal account will be given of the basic aspects of the analysis, interpretation and reporting of karyotypes in clinical cytogenetic work. It seems inappropriate here to give detailed instructions for the planned investigation of every possible case, since resources and workload are not evenly distributed between laboratories. By necessity each one must have its own guidelines. Some are nearly drowned in a torrent of amniotic fluids; others can afford the luxury of applying every possible technique to the interesting cases. But proper action is always crucial in clinical work and proper action cannot be taken unless one is aware of all the various dimensions the work may involve. This seems to be one of the most neglected aspects in other texts on human cytogenetics, and I will therefore try to concentrate upon it. Towards the end of the chapter, a few notes on laboratory data management, the use of chromosome repositories and quality assessment will be included.

2. CLASSIFICATION OF CHROMOSOMES

2.1 The Metaphase Chromosome

This may require some explanation for the newcomer to the field. The chromosomes as seen in a metaphase spread are tightly coiled packages of DNA and protein. They consist of two identical sister chromatids that would have separated at anaphase and ended up in separate daughter nuclei had they not been fixed. Consequently, they appear as double threads, or rods, held together at at least one point along their length, namely the primary constriction. This defines the position of the *centromere*, which is the attachment site of the spindle fibres. It divides the chromosome into a short arm and a long arm. The free end of an arm is called a *telomere*. A chromosome with its centromere in the middle is *metacentric* (e.g., chromosome 1). A chromosome with the centromere right at the end is *telocentric*. Telocentric chromosomes are not normally found in human cells. Those chromosomes with a centromere very near the end and with a very small short arm are *acrocentric*. Two groups of acrocentric chromosomes are present in the normal human cell. Their short arms are usually satellited. The satellite is a small knob of chromatin connected to the rest of the chromatid through a thin stalk, much like an antenna. The remainder of the chromosomes are classified as *submetacentric*. Their centromeric index (i.e., the ratio of the length of the short arm to the total length of the chromosome) may allow their classification within groups in a few cases.

Although the length of the chromosomes in human metaphase spreads is very much dependent upon the method of preparation, the average size of the longest chromosome (no. 1) is about 10 μm and the smallest (no. 21) is about 2 μm. The limit of optical resolution with light microscopy is about 0.2 μm. It must be realised that structures smaller than that are impossible to observe as separate entities, no matter how good the preparations are.

2.2 **Basic Chromosome Classification Scheme**

A conventional scheme for classifying the human chromosomes was developed in the early 1960s. This is summarised in the International System for Human Cytogenetic Nomenclature (1978) (1): 'When (human) chromosomes are stained by methods which do not produce bands, they can be arranged into seven readily distinguishable groups based on descending order of size and the position of the centromere:

Group 1−3 (A),	large metacentric (nos. 1 and 3) or submetacentric (no. 2) chromosomes readily distinguished from each other by size and centromere position;
Group 4−5 (B),	large submetacentric chromosomes which are difficult to distinguish from each other;
Group 6−12-X(C),	medium-sized submetacentric chromosomes. The X chromosome resembles the longer chromosomes in this group. This large group is the one which presents major difficulty in identification of individual chromosomes without the use of banding techniques;
Group 13−15(D),	medium-sized acrocentric chromosomes with satellites;
Group 16−18(E),	relatively short metacentric chromosomes (no. 16) or submetacentric chromosomes (nos. 17 and 18);
Group 19−20(F),	short metacentric chromosomes;
Group 21−22-Y(G),	short acrocentric chromosomes with satellites. The Y chromosome is similar to these chromosomes but bears no satellites.'

It is very rare these days that one would have to resort to unbanded preparations for the final analysis of chromosomes in clinical cytogenetics, but they are frequently used for preliminary screening because trypsin-Giemsa banding techniques require some ageing of the slides for optimal results. Generally accepted procedures for the actual analysis are outlined below in Section 4.4 describing direct analysis under the microscope.

2.3 **The Chromosome Constitution**

In the normal diploid cell with 46 chromosomes, one chromosome within each homologous pair represents the maternal inheritance and the other the paternal inheritance. Twenty two of the pairs are the same in males and females. They are called autosomes. The twenty third pair, the sex chromosomes, consist of two X chromosomes in the female and one X and one Y chromosome in the male. Abnormal cells with one extra chromosome are *trisomic* and those with one missing chromosome are *monosomic*. *Aneuploidy* is the collective term for numerical deviations of just one or a few chromosomes. If only a part of a chromosome is extra, or missing, the corresponding terms are partial trisomy and partial monosomy, respectively. Such cells are said to be genetically unbalanced. This is a much wider concept than aneuploidy which would not include partial deletions or duplications of a particular chromosome. *Triploid* cells have 69 chromosomes consisting of three haploid sets and *tetraploid* cells have 92 chromosomes. *Endoreduplicated* cells have chromosomes with more than two sister-chromatids, usually structurally normal. Structurally abnormal chromosomes are described below in the section on interpretation of abnormal karyotypes. A karyotype represents the total cytogenetic characteristics of the chromosome content of a single cell nucleus.

2.4 Identification of Banded Chromosomes and Chromosome Bands

Each chromosome in the human somatic cell complement can be uniquely identified following a number of different banding procedures as described in Chapter 3. The banding patterns are highly characteristic. Normally, it is not possible to distinguish by microscopy the maternally transmitted homologue from the paternal one, but there are exceptions to this rule. For the study of rearranged chromosomes, and in particular for the assignment of breakpoints, more detail can usually be seen with the finer bands present at late prophase which fuse together and form thicker bands in condensed metaphase chromosomes. This forms the basis of high resolution banding techniques. The ISCN (1981) (2) provides schematic representations, or ideograms, of chromosomes corresponding to approximately 400, 550 and 850 bands per haploid set. This band numbering system emerged from the Fourth International Congress of Human Genetics in Paris in 1971 and although under constant revision, seems to have been generally adopted. Clearly, there is a need for a uniform system of nomenclature. Its principles rest with a numbering system based on major bands as they appear from the centromere outward along each chromosome arm. The short arm is designated p, and the long arm is designated q. Let me now cite ISCN (1978) (1): 'A band is defined as that part of a chromosome which is clearly distinguishable from its adjacent segments by appearing darker or lighter with one or more banding techniques. Bands that stain dark with one method may stain lightly with other methods. The chromosomes are visualised as consisting of a continuous series of light and dark bands, so that, by definition, there are no interbands'. The centromere, telomeres and the mid-point of certain bands are called landmarks. Segments between landmarks are called regions. For example, the long arm of chromosome 1 consists of four regions numbered consecutively q1, q2, q3 and q4, from the centromere outwards. A band used as a landmark is considered as belonging entirely to the region distal to the landmark and, hence, it is band number 1 of the region distal to the landmark. Bands are then subdivided, which is indicated by a full stop, into sub-bands as they appear on more elongated chromosomes. For example, the fragile X locus has been assigned to band Xq27.3 (i.e., on the X chromosome's long arm region 2 band 7 sub-band 3).

For some reason ISCN (1978) (1) also allows another yardstick for specification of breakpoints. Each band is then divided into 10 units. 1p1206 would mean a point six tenths of the distance from the proximal edge of band 1p12. There must be very few cytogeneticists who can claim to have seen sharply defined breakpoints on this miniature scale.

3. APPLICATIONS OF STAINING AND BANDING TECHNIQUES

3.1 Solid Staining of Chromosomes

These techniques are frequently used for checking the quality of chromosome preparations and are also used for the preliminary screening in certain cases. Chromatid-type breakage and fragile Xs are not easily seen in banded preparations. Instant staining is obtained if the dye is dissolved in temporary mounting media. This can easily be washed off and the slide be used for various banding purposes. Solid Giemsa-staining can also be used although it is not as quick. Acetic orcein produces beautiful preparations that, unfortunately, cannot be used advantageously for any other purpose thereafter.

3.2 Q-banding

This was the first chromosome banding technique to be described, it was used as the reference for the Paris nomenclature [i.e., ISCN (1978) (1)], and it remains the simplest, most reliable and most informative of all the banding methods. However, it has several disadvantages as compared with most other banding techniques:

(i)   the preparations are not permanent;
(ii) the emitted light is weak and the staining fades quickly;
(iii) the contrast is low;
(iv) higher skills and better, and more expensive, equipment is required than with other techniques;
(v)   it is more labour intensive.

Therefore, it is not as suitable for routine diagnostic work on a large scale as G- or R-banding, although it remains a necessary complement to these methods. Its main use in our laboratory is for the instant diagnosis of specific abnormalities in urgent cases and also for the investigation of polymorphic variation in heterochromatic regions, which will be described in detail in Section 3.3.

3.3 Q-band Heteromorphism

3.3.1 *The Y Chromosome*

This shows an intensely fluorescent segment (i.e., band Yq12), on the distal part of the long arm. The fluorescent spot can also be seen in interphase nuclei, which provides an easy technique for sexing non-dividing cells, although it is less reliable than the examination of Barr bodies. Combining the two techniques is recommended. Fluorescent spots in the interphase nucleus do not always represent a Y chromosome. The heterochromatic area adjacent to the centromere, particularly on chromosomes 3 and 4, and the satellites of the acrocentric chromosomes 13−15 and 21−22 may show intensive fluorescence which can be confused with that of the Y chromosome. In a few cases, translocation of band Yq12 on to an autosome, most frequently to the short arm of chromosome 15, can be observed in normal females. Furthermore, band Yq12 is variable in size and the complete lack of this segment is seen in phenotypically normal males from time to time.

3.2.2 *Chromosomes 3, 4, 13−15 and 21−22*

Polymorphic variation of heterochromatin on chromosomes 3, 4, 13-15 and 21-22, as seen with Q-banding, produces useful markers and may frequently allow differentiation between the maternally and the paternally transmitted homologue within these pairs. This has proved to be a useful method, for example, for tracing the origin of the extra chromosome 21 in Down's syndrome. Examination of the full pattern of such markers in triploid fetuses, hydatidiform moles and ovarian teratomas has also been of scientific value. In clinical practice, it is probably the simplest way to distinguish between true fetal XX/XY mosaicism and maternal cell contamination of amniotic fluid samples, but it does not always work and it requires considerable experience, since there are many pitfalls.

3.3.3 *Problematic Cases*

Although most cytogeneticists seem to realise the technical difficulties associated with the examination of heteromorphic variation, very few seem to appreciate the high frequency of new mutations in heterochromatic segments. This may have dangerous consequences. For example, the finding of a boy having a smaller Y chromosome, with a smaller fluorescent segment than his father, must not under any circumstances be taken as sufficient evidence that the biological father might be someone else.

3.4 **G-banding**

These are the most popular techniques for chromosome banding in routine cytogenetics laboratories. Trypsin-Giemsa is most commonly used. The banding patterns produced are very similar to Q-bands, except for the heterochromatic segments. The large blocks of heterochromatin adjacent to the centromeres on chromosomes 1, 9 and 16 which show very dull fluorescence with quinacrine are variable in their staining reaction with Giemsa. Those on chromosomes 1 and 16 usually stain intensely; the one on chromosome 9 is usually pale. Satellites are less informative than with Q-banding and the distal part of the Y chromosome is usually stained but unremarkable. The great advantage of these techniques is that they produce reasonably permanent slides with good contrast. No equipment other than a good microscope is needed for the analysis, but the techniques are not without their problems. They all require some pre-treatment of the preparation which may destroy it. Over-trypsinisation is a frequent problem. Furthermore, most procedures require some ageing of the slides for optimal results. The precise problem is that all procedures involving acid and basic dyes in combination like Giemsa and other Romanowsky stains are difficult to standardise. The chromosome banding is entirely dependent upon the inherent tendency of these stains to sharply differentiate between virtually identical chemical structures and the formation of precipitates; usually on the surface of the chromatid.

It will be assumed for the rest of this chapter that trypsin-Giemsa is the standard method. Should your laboratory use Q-bands or R-bands, the section on microscopic analysis may not apply directly, but the principles will certainly remain the same.

3.5 **R-banding**

Q- and G-negative bands appear positively stained with R-banding techniques and *vice versa*. Heterochromatic regions generally remain unstained with R-banding techniques, with the exception of telomeres which are better visualised than with other methods. On the other hand, it could be argued that the ends of chromosome arms are usually sharply defined with trypsin-Giemsa, even if negatively stained. Reverse banding is sometimes very useful especially for the characterisation of rearrangements involving the smaller F- and G-group chromosomes 19−22. It is also recommended that, for example, reciprocal translocations are studied with both G- and R-banding techniques for the localisation of breakpoints, although this is rarely considered necessary in clinical practice. The disadvantages with most R-banding techniques are that they usually do not produce as high contrast as G-banding techniques. Phase contrast microscopy is frequently required, and fluorescence microscopy has to be resorted to for the two R-banding techniques described in Chapter 3. Although rarely used in our laboratory, R-banding is considered the method of choice for routine work in some other laboratories.

3.6 **BrdU Techniques**

The substitution of 5-bromo-2'-deoxyuridine (BrdU) for thymidine in DNA dramatically alters the staining characteristics of chromatin. This can usefully be exploited for visualising G-bands, R-bands, T-bands, sister chromatid exchange (SCE), and the late replicating inactive X chromosome in female cells. Some high resolution techniques are based on BrdU substitution as well as on the interference of this drug with DNA replication. The banding patterns obtained are often excellent and highly reliable, provided a perfect timing of the BrdU pulse is achieved in relation to the cell cycle. The disadvantage of these techniques, as far as analysis is concerned, is the increased background level of spontaneous aberrations, which rarely matters very much, and also the fact that it might be difficult to elicit other banding patterns than those for which the preparations were designed. Furthermore, with some of these techniques, G-banded metaphase spreads may appear together with R-banded spreads and SCE patterns on the same slide. This may be confusing. However, a more serious disadvantage seems to be that they are more labour intensive, and also require a precision in timing that is not always appreciated. For those who are prepared to take the trouble the results are often rewarding. Only a few applications will be mentioned here. G-bands or R-bands for ordinary routine work as well as for high resolution studies will not be commented upon further.

(i)   A special application of these techniques is for the preparation of slides for *in situ* DNA hybridisation (see Chapter 7). It is often difficult to obtain good banding of chromosomes subjected to denaturation procedures and buried in photographic emulsion unless BrdU techniques are used.

(ii)   The application of BrdU techniques for visualising the late replicating X chromosome in female cells is based on the same principles as classical autoradiography. BrdU techniques are far more convenient. Any human chromosome can be regarded as a linear array of independent replicating units (replicons). The time of onset of DNA synthesis in each unit is regulated by an as yet unknown mechanism. Corresponding replicons on homologous chromosomes usually initiate replication at the same time. The late replicating inactive X in female cells is an exception to this rule. Its replicons are timed later than their counterparts on the active X. However, chromosomal rearrangements may upset the X-inactivation pattern (see Section 10.5). For example, in X-autosome translocations only the derivative chromosome receiving the X-inactivation centre (located at Xq13) can be expected to be inactivated. These cell lines are usually lost already during embryogenesis but can be observed in the exceptional case. Spreading of X-inactivation onto the autosomal segment may occur. X-inactivation studies should be carried out in such cases if one wants to correlate the phenotype with the chromosomal findings.

(iii) SCE studies have applications for the study of chromosome breakage syndromes. It can also be used for mutagenicity testing although this is rather unreliable and rarely done in routine clinical work.

3.7 C-banding

There are at least two applications of C-banding techniques in clinical cytogenetics. Sequential staining of the same cells, after previous identification of the chromosomes with, for example, Q-banding, is usually required. One application is for the study of polymorphic markers. The paternal and the maternal homologue can sometimes be distinguished, especially for chromosomes 1, 3, 4, 9, 13−15, 21−22. The other application is for rearranged chromosomes. It may be interesting to see to what extent centromere heterochromatin is involved in the rearrangement. Supernumerary small marker chromosomes should always be studied by this technique to complement Q-banding studies.

3.8 NOR Staining

The main application of this technique is for the study of satellited chromosomes. Satellites on chromosome arms other than the short arms of 13−15 and 21−22, should always be studied. Satellited small extra markers are the most frequent indication. NOR staining is also useful for the study of polymorphisms in combination with other banding techniques.

3.9 DAPI/Distamycin A Staining

This is especially useful for the study of anomalies near the centromere on chromosome 15. A substantial percentage of small accessory chromosomes are thought to originate from this region.

3.10 G-11-banding

Alkaline Giemsa is sometimes used to differentiate between mouse and human chromosome material in cell hybrids (see Chapter 7). It can also be used for distinguishing between unsubstituted and BrdU-substituted chromatin. BrdU-substituted chromatin will show more intense staining than unsubstituted chromatin which stains very lightly, with the exception of the variable region on chromosome 9.

3.11 Sex Chromatin

Although frequently used in the past, these methods have very limited application in current practice. As a rule, chromosome studies should be done instead whenever possible. They are far more informative and nearly 100% reliable. Many cytogeneticists today have limited experience of screening for Barr bodies, drum-sticks and fluorescent Y spots.

3.12 Other Staining and Banding Techniques

Other techniques are occasionally also applied in clinical cytogenetics. However, most of these are variants of those described here. Completely new banding patterns have recently been demonstrated. For example, nick translation *in situ* following treatment with various restriction endonucleases produces patterns that may find clinical applications in the future.

4. DIRECT ANALYSIS UNDER THE MICROSCOPE

4.1 Problems of the Human Mind

The analysis of chromosomes in clinical routine work is usually carried out directly under the microscope. It takes a great deal of training before anyone without experience is able to find 'good' cells, to count and classify chromosomes correctly, and to disregard findings of an artefactual nature. Most learners like to resort to some kind of illustration of the cell with which they are struggling. A drawing prism fitted to the microscope is very useful for teaching purposes. Photographic prints or polaroid pictures can also be used since they accurately represent the view. However, photographs are always slightly inferior compared with looking through the eyepieces; their main use is for documentation purposes. All aberrant karyotypes should be properly documented. Assuming that all the details of microscopy in Chapter 3 have been digested, I will only make a few points here regarding the analysis.

4.2 Setting up the Microscope

The first point concerns contrast. The human eye has an optimum discriminatory power for green light. The commonly used Giemsa stain produces pink, red, purple or blue staining of chromosomes. This means that the least strain to the eyes and maximum contrast is usually obtained with a green filter. Should the contrast be too high, one could use a light blue filter, or no filter at all. Phase contrast microscopy intrinsically gives a lower resolution and is also more time-consuming, but may be used as a last resort if the staining is too pale.

4.3 Locating Metaphase Spreads Suitable for Analysis

Finding the best cells on a microscope slide is a matter of skill and hard work. Systematic screening under a low power objective is the only way that is efficient in the long run. Do not start off in the far corner because cells tend to remain fairly close to where the suspension was applied on the slide. The best cells to analyse usually have between 400 and 600 bands per haploid karyotype [ISCN (1981) (2)], depending on the indications for the analysis. The ACC panel of advisors in the UK quality assessment scheme, 1984, used the following criteria for a good cell: four bands on the 18q, three bands on the 11p, and a good distinct band on 17p. The 15q12 band should be clearly seen in a very good cell and there should also be two bands on 20p. Very elongated chromosomes are unduly time-consuming, even for the most proficient analysts. They also present a serious problem of homologue discordance. The actual banding technique used is usually not important. Most cytogeneticists tend to favour the banding technique with which they have become proficient in analysis. It is a frustrating experience to realise that it takes a great deal of effort to learn, for example, the reverse banding pattern, even if one knows the ordinary trypsin-Giemsa pattern perfectly well.

4.4 Analysis

The analysis of a cell is a systematic process including a band-for-band comparison of homologues.

Reference to the location of a particular cell is made by writing down the coordinate readings of the microscope on the analysis sheet before the actual analysis. Unfortunately, the transfer of coordinates between different microscopes is rather cumbersome. One can use an England finder. Make sure it is clear to which microscope the readings refer. Documentation is crucial.

Reference to the location of a particular chromosome in a spread is conveniently made with the help of an imaginary grid of a clock face centred over the group of chromosomes in view. For example, '4 o'clock 2/3 from the centre towards the periphery' would in most cases be accurate enough to convey the idea of which chromosome is under discussion.

It is best to start the analysis by counting the chromosomes. Subdividing the spread into two or three sections usually makes it easier. Bear in mind that satellite associations of the acrocentric chromosomes might be misleading. The further analysis of unbanded spreads would normally proceed as follows.

(i)   Locate the A group chromosomes 1−3 and compare the size and centromeric index of homologues.
(ii)   Look next at the B group chromosomes 4−5, which are indistinguishable from each other.
(iii) Switch to the smallest acrocentrics, the G group 21−22, including the Y chromosome, which is sometimes distinct.
(iv)   Work backwards with the smallest metacentrics, the F group 19−20.
(v)   Proceed to the submetacentric E group 16−18, where discrimination is usually possible.
(vi)   Look next at the larger acrocentric chromosomes, the D group 13−15.
(vii) Finally, the remaining C group submetacentric chromosomes 6−12, including the X, are inspected and counted: 15 in the normal male, and 16 in the normal female.

In banded preparations, each chromosome should be individually located. A detailed comparison of banding patterns between the two homologues is mandatory. Telomeres are especially important. Any suspected discrepancy should be followed up in a number of consecutive cells until a decision can be made as to whether one of the two homologues is abnormal or not.

An example of how the analysis results could be presented on the analysis sheet is as follows: '85B/394-1; 12.8/85.3; 46,XX; all analysed/both 6s OK', where the first group of characters would uniquely identify the slide, the second group the coordinates, the third group a normal female karyotype, 46,XX and the fourth group a comment which indicates that all chromosomes have had their banding pattern examined, and that the findings contradict a suspicion from a previous cell that there might be something odd about one chromosome 6.

4.5 Number of Cells to be Analysed

How many cells should be analysed is a more difficult question to answer. Although the pitfalls connected with analysing too few cells are fairly obvious, the merits of analysing too many are questionable. It depends entirely on the indication, the quality of the preparations and the skill of the microscopist. In most cases, he or she has a preconceived idea of how many cells ought to be analysed. Two cells are usually enough to rule out the possibility that someone is a carrier of a familial rearranged chromosome. Ten cells are adequate for most purposes. Thirty cells are needed to rule out a $10-15\%$ or more mosaicism. Two hundred cells are usually regarded as the minimum to rule out a low grade mosaicism. It is rarely necessary to fully analyse more than three of the best cells if the preparations are good. Poor quality preparations will never give the required answer, no matter how many cells are analysed.

4.6 Acquired Chromosome Aberrations

Acquired chromosome abberations may appear in any preparation. There is always a low background level of spontaneous chromosome breakage and rearrangement in normal cells. This level may vary, not only between individuals but also between laboratories having different techniques for growing cells and making the preparations. On the whole, it remains low and usually does not interfere with the analysis of constitutive karyotypes. However, the appearance of clones of abnormal cells may sometimes present difficulties of interpretation, especially in prenatal diagnosis (see Section 12.4) and, to a lesser extent, in pre-malignant disorders (see Section 11.2.4).

Induced chromosome breakage in normal subjects can be found following exposure to a number of toxic agents, radiation and viral infections. Radiation-induced damage may be demonstrated years after the exposure as, for example, by a relatively high frequency of dicentrics in phytohaemagglutin (PHA)-stimulated lymphocytes, but dose-response relationships for the other mutagens are usually very unpredictable. In this context one must not forget that the aberrations may also be due to some very rare condition that shows a high frequency of spontaneous chromosome aberrations or, alternatively, a high susceptibilty to induced damage. Such are the chromosome breakage syndromes, including Bloom's syndrome, ataxia telangiectasia and Fanconi's anaemia: they are all associated with a high incidence of malignancy (Section 11.2.4).

4.7 **Artefacts**

Genuine artefacts are to be disregarded. The ugliest ones I know are over-trypsinised chromosomes and chromosomes that have been scratched by filter paper.

5. KARYOTYPING

5.1 **Definition and Use**

A karyotype defines the total characteristics of the chromosomes from a cell nucleus as seen with the microscope. 'Karyotyping' usually means the process of cutting out the chromosomes from photographic prints of a cell, and sticking them on to a piece of cardboard, arranging them according to size and form. This makes a comparison of banding patterns of homologues easy and it is also possible to compare the individual chromosome with standard ideograms [e.g., ISCN (1981) (2)]. Partial karyotypes of, for example, the two chromosomes involved in a reciprocal translocation, together with their normal homologues, are also frequently used for documentation. Some laboratories routinely prepare karyotypes from all cases. Others do not produce them at all. They are time-consuming, and probably not meaningful in routine cases unless there is a more complex rearrangement that needs illustration. The documentation of reciprocal translocations by partial karyotypes is sufficient in most cases.

5.2 **High Resolution Studies**

Complete karyotypic analysis of prometaphase chromosomes in high resolution studies is virtually impossible to do directly under the microscope and usually requires a set of karyotypes prepared from cells with a similar amount of chromosome contraction. Very elongated chromosomes at, say, the 800 band level, would require up to as many as 10 karyotypes, due to homologue discordance in the banding patterns, overlapping chromosomes, and other additional influences (3). This sort of effort is impractical for routine work, which means that high resolution studies are only indicated for specific chromosomes, for which partial karyotypes should be prepared, rather than for routine screening of complete karyotypes.

6. DETERMINATION OF BREAKPOINTS

6.1 **Exchange Aberrations and Pattern Disruption Points**

The concept of a breakpoint has a flavour of exactness that for several reasons is rarely achieved in cytogenetics. On the contrary, a breakpoint as given in the description of a chromosome rearrangement often represents a feeling of likelihood rather than being a claim that it is the only possibility. Any chromosome, even if it is as sharply banded as a coral snake, could easily exchange the distal part of an arm with another chromosome having a similar banding pattern on the end, without much noticeable effect, unless one was cut shorter than the other. Even then it might be impossible to tell in which band the break occurred, because the banding pattern might still match with the normal homologue. In theory, there is an inherent three-band uncertainty since pattern disruption points do not necessarily coincide with the breakpoints (4). In practice, an assignment of the breakpoint can usually be made with reasonable certainty, but not in every case. The banding patterns rarely merge completely and, if they do, the rearranged chromosome can be studied by different techniques which may reveal subtle differences in band size, staining intensity and conformation, as compared with the normal homologue. High resolution banding studies should always be considered. In the case of unbalanced offspring, study of the same translocation in its balanced form in a parent may give valuable further information.

6.2 **Non-exchange Aberrations**

Non-exchange aberrations such as simple terminal deletions, chromatid breaks and so on are easier and allow for a band-for-band comparison with the normal homologue. It is recommended that breakpoints are reported according to ISCN (1981) (2).

7. DESCRIBING AN ABNORMAL KARYOTYPE, PARIS NOMENCLATURE

7.1 **ISCN (1978)**

The purpose of the following paragraphs is not to recapitulate the International System for Human Cytogenetic Nomenclature (1978) (1) since it is assumed that every reader of this chapter has a copy available for easy reference. My intentions are rather to provide a few arguments on how to best use it in commonly encountered situations in clinical cytogenetics. In this context I also feel obliged to point out that ISCN (1978) (1) includes some rather artificial structures formed from unrelated dialects in cytogenetics which may lead the student into complete confusion from time to time. For simplicity, the discussion will be strictly limited to constitutional aberrations seen in banded preparations.

7.2 **Cytogenetic Nomenclature**

The meaning of some symbols and abbreviated terms is worth a comment. Let us start with terminal deletions. For example, a break at band 1q21 and deletion of the long arm segment distal to it would intuitively be thought to be described by this nomenclature as del(1)(q21 → qter). However, the correct description should be either del(1)(q21)(sic!) or del(1)(pter → q21:), which actually describes the segment that was not deleted. Do we need this elaborate confusion? I do not think so and will use the intuitive notation under these circumstances. The double colon indicates breakages and reunion of bands. For example, an interstitial deletion between 1q21 and 1q31, that is del(1)(q21 → q31) in our intuitive notation, should correctly be written del(1)(pter → q21::q31 → qter) or del(1)(q21q31), without comma, semicolon, colon, double colon, arrow, full stop or space. Had band 1q21 been specified as 1q21.3 the short notation would have been del(1)(q21.3q31). The very innocent might wonder what role chromosome 3 has in this deletion. Having said that, it gives some comfort to find that a correct short notation for an inverted duplication would be inv dup(2)(p23 → p14) which is both unambiguous and redundant, since the breakpoint more proximal to the centromere is always specified first when the two breaks occur within the same arm.

The strength of ISCN (1978) (1) lies in the 'short' system for reciprocal translocations and their unbalanced products using the *der* symbol which is efficient, except for the redundant brackets. For example, 47,XX, + der(5),t(2;5)(q21;q31)mat is a full story (see *Figure 4*). It is probably worth pointing out that the specification of the translocation and the fact that it came from the mother needs to be written out only once in any report. It will always be understood that der(2) means the rearranged chromosome having the centromere from chromosome 2 and *vice versa*.

Robertsonian translocations (*Figure 3*) can be written in at least six different ways. It appears that t(13q14q) is the most efficient notation. Other whole-arm translocations are described slightly differently, and, for example, t(2;3)(2p3p;2q3q) although unambiguous is redundant, since the first parenthesis does not contain any information that is not contained in the second one.

A pericentric inversion of chromosome 2 with the breakpoints at bands 2p13 and 2q24 is correctly written inv(2)(p13q24). I do not not think that anybody would actually complain if one wrote inv(2)(p13→q24) which is more explanatory. The recombinants 1 and 2 in *Figure 2* would be described as rec(2)dup p and rec(2)dup q, respectively.

As regards isochromosomes, I must admit that I always report, for example, 46,X,isoXq rather than i(Xq) because the clinician usually understands what is meant by *iso* whereas the letter *i* does not evoke any associations in his or her mind, unless it is a capital letter.

With three-break rearrangements the notation again becomes rather convoluted. For example, dir ins(2)(p13q21q31) specifies an insertion of 2q21→2q31 into 2p13. One might wonder whether dir ins(2)(p13,q21→q31) would have been clearer. Unfortunately, it is hopelessly incorrect. Just think of the comma. Had the segment come from another chromosome the correct notation would have been dir ins(5;2)(p14;q22q32) for the transfer of 2q22→2q32 into 5p14 with band 2q22 nearest the centromere. Any secretary will immediately spot the asymmetry within the second parenthesis and either remove the first semicolon or fill in the second one which is presumably missing.

Similarly, the complex translocation t(2;5;7)(p21;q23;q22) is uniquely specified by the order of the appearance of chromosomes within the first pair of brackets which might be difficult to appreciate. Fortunately rearrangements involving more than two breakpoints are very rare.

7.3 Reporting a Karyotype (ISCN 1978)

Having achieved proficiency in Paris nomenclature [i.e., ISCN (1978) (1)] do not try to excel. I do not think it would help the clinician very much if the chromosome report on a newborn baby was formulated as:

47,XY,+21,var(21)(p13,Q12),var(21=2)(p13,Q54)mat

The chances are that he or she might think that this represents a most extraordinary case of Down's syndrome.

The important information is that this represents a regular trisomy 21 in a male. Similarly, unbalanced karyotypes like 46,XX,−5,+der(5), t(2;5)(q21;q31)mat should be explained in plain words as, for example, 'abnormal unbalanced female karyotype with partial monosomy for the distal part of the long arm of chromosome 5 and partial trisomy for the distal part of the long arm of chromosome 2. The mother is a carrier of the same translocation in its balanced form, and might be at risk of having further children with the same abnormality'. This will enable the clinician to take proper action and also to gain access to similar cases from the literature. Furthermore, it will not matter very much if the maze of ISCN (1978) (1) nomenclature has deluded the cytogeneticist.

8. REASON FOR REFERRAL AND RATE OF DETECTION OF CHROMOSOME ANOMALY

8.1 Family History

More than 2/3 of chromosomal errors associated with early spontaneous abortion and nearly half of those detected among newborns involve an extra chromosome. For most of these there is no family history and the recurrence risk is really quite low in the younger mothers ($\sim 1\%$) (5). On the other hand, certain chromosomal rearrangements can easily be traced through generations because of their tendency to produce recurrent miscarriage and birth defects. *Figure 1* illustrates, as an example, a family with a reciprocal translocation t(14;21)(q32.1;q22.2). The anomaly was missed at the first chromosome investigation because the proband with Down's syndrome was thought to represent a regular trisomy 21. The predigree implies that the mother's youngest brother is also likely to be a translocation carrier because his wife has had five successive spontaneous abortions. Unfortunately, not all clinicians realise the importance of conveying this sort of information to the cytogenetics laboratory. Typically, the chromosome request card would say '? Down's'.

8.2 Fetal Chromosome Analysis on Grounds of Maternal Age and/or Low Maternal Serum AFP

Chromosome aberrations occur in 0.6% of liveborn children and the majority of these aberrations cause severe handicap. A minor proportion of the affected pregnancies can be identified in time to give the parents the option of selective termination. Apart from those with a previous child with a chromosome aberration, and known carriers of chromosome rearrangements, this prenatal diagnostic service is mainly used on the grounds of maternal age being 35 years and over. The risk concerns aneuploidy. Rates increase exponentially with advancing maternal age for trisomies 21, 18 and 13, and for the 47,XXX and 47,XXY karyotypes. The maternal age-specific rates for trisomy 21 are dealt with in Section 12.1 on prenatal diagnosis and screening for Down's syndrome (*Table 3*).

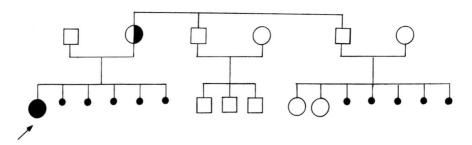

Figure 1. A family with a reciprocal translocation t(14;21)(q32.1;q22.2) in a mother and child with 47,XX, −14, +der(14), +der(21) Down's syndrome.

It has recently been shown that Down's syndrome fetuses are associated with a lower level of maternal serum alpha fetoprotein (AFP) as well as a lower level of amniotic fluid AFP than normal. This could be used in combination with the maternal age factor to improve the prediction as to which pregnancies are at risk, especially in the 30−35 years of age group of women (6).

8.3 Newborns, Children and Perinatally Abnormal Infants with Growth Retardation and Malformations, or Dysmorphic Features Suggesting a Chromosome Anomaly

Although the overall rate of chromosome aberrations in unselected newborns is about 0.6% (*Table 1*) the rate in selected groups is very much higher and entirely dependent on the paediatrician's individual skill in recognising chromosome syndromes. Perinatally abnormal infants show chromosome abnormalities in about 5−6%.

8.4 Mentally-retarded in Institutions

Again the selection criteria are most important, and 10−30% are not unusual figures. Down's syndrome patients tend to account for 10−15% of the mentally retarded in institutions; X-linked mental retardation is also common.

8.5 Infertility and Recurrent Abortions

Infertile males with normal sperm counts are unlikely to have a chromosome aberration (0.2%). On the other hand, 1/7 of those who are aspermic show chromosome aberrations, mostly sex chromosome aberrations (e.g., 47,XXY Klinefelter's syn-

Table 1. Detection of Chromosome Anomaly at Amniocentesis in Pregnancies of Women ≥ 35 Years of Age Compared with Rates Determined from Surveys of the Newborn.

Type of chromosome abnormality	Prenatal diagnoses (%)	Newborn surveys including all maternal ages (%)	Fraction
Autosomal anomaly			
Trisomy 21	1.16	0.12	1 in 700
Trisomy 18	0.23	0.01	1 in 3000
Trisomy 13	0.07	0.01	1 in 5000
t(13q14q)	0.05	0.07	
Other balanced structural rearrangements	0.18	0.12	
Extra marker chromosome	0.06	0.02	
Other unbalanced structural rearrangements	0.08	0.06	
Sex chromosome anomaly			
47,XXX	0.25[a]	0.10[a]	1 in 800[a]
47,XXY	0.33[b]	0.09[b]	1 in 700[b]
47,XYY	0.07[b]	0.09[b]	1 in 800[b]
45,X	0.09[a]	0.01[a]	1 in 2500[a]
Other unbalanced	0.05	0.06	

[a]Rates for females.
[b]Rates for males.
Data from the European Collaborative Study (10) and (16).

Table 2. List of Examples of Reason for Referral (excluding queries for specific chromosome anomaly).

Problem	Look for anomalies involving chromosome and/or band No.
Ambiguous genitalia	X and Y but ACCBI[a]
Amenorrhea	X
Anaemia and mental retardation	16p
Aniridia	11p13
Azoospermia	X and Y but ACCBI
Behaviour problems	X and Y but ACCBI
Cat-Eye syndrome	22(pter-q11)
Cleft lip and palate	ACCBI
Congenital heart disease	ACCBI
Developmental delay	ACCBI
Di George syndrome	22q11
Epilepsy	ACCBI
Exomphalos	18 but ACCBI
Failure to thrive	ACCBI
Floppy baby	ACCBI but check P-W band 15q12
Gynaecomastia	X or Y
Hypospadia	ACCBI
Infertility	X or Y but ACCBI
Langer Giedions syndrome	8(q22-q24)
Macroorchidism	fra(X)
Mental retardation	X but ACCBI
Microcephaly	ACCBI but check 4p and 5p in particular
Multiple abnormalities	ACCBI
Multiple endocrine neoplasia	20p12
Odd facies	ACCBI
Prader-Willi syndrome	15(q11-q12)
Recurrent miscarriages	ACCBI
Retinoblastoma	13q14
Small stature	X
Tall	X or Y
Wilms tumour	11p13

Some data from (18).
[a] Any Chromosome Could Be Involved = ACCBI.

drome). Similarly, primary or secondary amenorrhoea is not infrequently caused by X chromosome aberrations; 45,X Turner's syndrome being the classic example. Couples with recurrent abortions show chromosome abnormalities in about 7–9% of cases (7).

8.6 Products of Conception

The incidence of chromosomally unbalanced early conceptuses is very high, but the finding of an anomaly is usually of very little predictive value for further pregnancies. The current policy in our laboratory when such specimens are received from first trimester pregnancies is to ask the clinician if we could analyse bloods from both parents instead.

A list of common reasons for referral and what specifically to look for is included in *Table 2*.

9. INTERPRETATION OF NORMAL KARYOTYPES AND THOSE CONTAINING A BALANCED REARRANGEMENT

9.1 Doubts About a Normal Analysis Result

Apparently normal chromosomes are the most common result in cytogenetic analysis and are therefore by definition usually right. However, if the clinician insists that there must be an underlying chromosome abnormality to account for the clinical features, one should always be prepared to repeat the investigation. There is always a possibility of a mix-up of samples. It is doubtful whether it is worth initiating high resolution studies unless a specific anomaly is suspected; these techniques are far too time-consuming to be of any practical value in general screening. Look at the parents instead! It might be easier to see a rearrangement in its balanced form. Make sure the appropriate investigations have been done. For example, it is of no use reporting a normal karyotype on ordinary lymphocyte cultures in cases of suspected fragile X syndrome when low-folate media should have been used instead (see Chapter 2). There are sometimes genuine difficulties in interpreting a 'normal' karyotype. This is usually in one way or another related to polymorphic variation of the individual chromosomes. The finding of an apparently balanced rearrangement in a patient with clinical features that might suggest a chromosome anomaly will also be discussed in this context.

9.2 Chromosome Polymorphisms Involving Heterochromatic Regions

Aberrations involving the heterochromatic regions on the long arm of the chromosomes 1, 9 and 16, and also on the short arm of chromosomes $13-15$ and $21-22$, and on the distal part of the long arm of the Y chromosome as well as centromeric regions on other chromosomes, can usually be regarded as normal variants. For example, the finding of the whole of the fluorescent region of the Y chromosome translocated onto the short arm of chromosome 15 in a female may seen rather alarming, but studies on several families with such translocations have not provided evidence for any effect on the phenotype, unless euchromatic portions of the Y chromosome have also been translocated. Similarly, satellites attached to the ends of non-acrocentric chromosomes, or double satellites on an acrocentric, are usually of no clinical significance.

9.3 Commonly Observed Pericentric Inversions, Fragile Sites and Other Variations Without Effect on the Phenotype

The very commonly found pericentric inversion of chromosome 9 moving the heterochromatic region from the long arm to the short arm is apparently without clinical significance and should be regarded as a normal variant. A small pericentric inversion on chromosome 2 is also regularly found and does not appear to have any effect whatsoever on the phenotype. This also applies to other small pericentric inversions on the larger chromosomes that are found from time to time, especially when analysis is carried out with high resolution techniques. These techniques have a notorious tendency to provoke homologue discordance in banding patterns which have to be distinguished from real variation which is sometimes very different.

Apart from the well known fragile site in band Xq27.3, linked to the Martin-Bell or fragile X syndrome, there is as yet little evidence that other fragile sites in the human karyotype are deleterious. A fragile site on 16q might be responsible for recurrent abortions. There are also other examples of unconfirmed deleterious effects.

9.4 Apparently Balanced Rearrangements in Patients with Clinical Findings that Might Suggest a Chromosome Anomaly

In a small number of these cases the simultaneous occurrence of abnormalities and an apparently balanced chromosome rearrangement will be coincidental, but there are reasons to believe that in the majority of the cases there may be an unidentified link between the rearrangement and the abnormal phenotype. There are several possibilities.

(i)   The rearrangement is more complex than it seems to be. There are several examples in the literature of apparently balanced reciprocal translocations that later, upon further investigation, have proved to be insertions or complex translocations involving three or more chromosomes. High resolution banding, meiotic studies and molecular DNA studies have provided the necessary evidence.

(ii) The breakpoints might have damaged essential gene function by splitting the gene or by position effects. There are several examples of this in the literature.

(iii) If the same rearrangement is found in one parent who is normal, further subtle rearrangements might have occurred during meiosis, for example, through unequal crossing over. This can usually be demonstrated with DNA techniques only. The phenomenon might be more common than the actual number of cases where this has been demonstrated would suggest.

(iv) The patient might have an undetected mosaicism involving an unbalanced cell line. Alternatively, the patient might be an unbalanced segregation product of an undetected mosaicism in one of the parents who either show a normal or a balanced rearranged karyotype. Examples of both these alternatives are known.

9.5 In Summary

Make sure the appropriate investigations have been done. There will always remain a proportion of unexplained cases which hopefully will gradually disappear as new techniques become available.

10. NOTES ON THE INTERPRETATION OF UNBALANCED KARYOTYPES

10.1 Clinical Findings in Autosomal Chromosome Aberrations: Principles and a Few Examples

The phenotypic effects of an autosomal chromosome abnormality are usually related to disturbances of gene dosage and regulating functions, rather than to structural defects in the individual genes. The fundamental concept is imbalance. Any deviation from the normal diploid karyotype is potentially hazardous to normal development. Autosomal imbalance usually has very serious consequences. Growth retardation, dysmorphic features, multiple malformations and mental deficiency are the most common findings. Very rarely would individual features be pathognomonic for a particular chromosome abnormality. The total pattern of abnormalities is frequently diagnostic. Such patterns of features are called syndromes. A chromosomal syndrome is a combination of characteristic dysmorphic features and malformations, mental retardation, and so on, caused by a specific chromosome aberration. The most important ones are briefly described in this section. Chromosome studies are mandatory to confirm the diagnosis even in typical cases, since the phenotype does not define the genotype unambiguously. The underlying condition might well be an unbalanced rearrangement, which usually implies a much higher recurrence risk than a simple trisomy. For example, Down's syndrome associated with mental retardation, characteristic facial features, palm print abnormalities and a variety of other useful diagnostic signs, none of which is entirely specific for this disease, can only be unequivocally diagnosed by a finding of trisomy 21, or at least extra material corresponding to band 21q22. The variation between individuals with the same chromosomal syndrome is usually relatively large but on the whole they can often be said to resemble each other more than they resemble their normal siblings.

Not all chromosomal syndromes involve a whole chromosome. Some phenotypes can be mapped to specific chromosomal segments: del(5)(pter→p15) defines the cri du chat syndrome described in the early days of clinical cytogenetics. This is one of the most frequently occurring autosomal deletion syndromes. It is characterised by microcephaly, severe mental retardation, a mewing cry almost one octave higher than normal, round face, hypertelorism, and so on. Again, none of the clinical findings are entirely pathognomonic but the pattern of features will make it easily recognisable.

The new banding techniques have allowed the characterisation of even more subtle deletions of chromosomes which have been found to be associated with clinical syndromes. For example, the Prader-Willi syndrome, characterised by floppiness, small genitalia, obesity and mild to severe mental retardation, has been found to be associated with an interstitial deletion of band 15q11 or 15q12 in about half of the cases. Discordance of the banding pattern of homologous chromosomes after high resolution banding procedures makes this small deletion of a chromosome a rather difficult diagnosis which hopefully in the future will be supplemented with molecular DNA investigations for confirmation.

Predicting a Prader-Willi syndrome phenotype from an apparent deletion of 15q12 is yet of doubtful value.

10.2 Autosomal Numerical Aberrations

10.2.1 Incidence of Aneuploidy

The incidence of aneuploidy at conception is thought to be about 10%, maybe even higher. Due to spontaneous abortion and perinatal mortality, the vast majority of chromosomally abnormal conceptuses are eliminated. Only three autosomal aneuploids survive fairly commonly till after term, namely trisomy 13, trisomy 18 and trisomy 21. The incidence of these disorders is about 1 in 5000 livebirths, 1 in 3000 and 1 in 700, respectively. The incidence of all three conditions increases with maternal age. Only those with trisomy 21 have a reasonable chance to survive into adult life. Well over 90% of the others die within a year of birth. Some other autosomal aneuploids occasionally survive to term and especially so when present in the mosaic condition.

Mosaic trisomy 8 is a chromosome disorder with a distinct clinical picture and long-term survival. Mosaic trisomy 2, 7, 9, 20 and 22 have also been reported occasionally, but are usually more detrimental.

10.2.2 *Trisomy 21, Down's Syndrome or Mongolism*

Down's syndrome has a highly characteristic pattern of features, yet there are sometimes difficulties in making the correct diagnosis in the newborn. Signs in the newborn include hypotonia, brachycephaly, flattened facial profile, epicanthic folds, Brushfield spots on the iris, protruding tongue, short and broad hands with short 5th finger and simian crease; fingerprint patterns are characteristic, increased space between the 1st and 2nd toe and so on. Birth defects are common. Congenital heart defects in about one third of the cases, and susceptibility to respiratory infections and childhood leukaemia account for a high death rate in the early years of life. Later on moderate to severe mental retardation is the most significant disability. It is the commonest known genetic cause of severe mental retardation, and accounts for $10-15\%$ of all mentally retarded in institutions. Some 95% of all patients with Down's syndrome have a $47,+21$ karyotype. One to two per cent are mosaic, having both $47,+21$ and 46 chromosome cell lines. The remaining $3-4\%$ have unbalanced rearrangements, mostly Robertsonian translocations, of which one half have arisen *de novo*: t(14q21q) and t(21q22q) are the commonest. Almost all of the homologous t(21q21q) unbalanced forms, which are rarely seen, are new mutations, Reciprocal translocations and other rearrangements involving chromosome 21 are also seen occasionally.

Mosaic Down's syndrome may involve less severe mental deficiency, and very exceptionally, mental development is normal, despite other typical features. The interpretation of mosaicism is always difficult. Sometimes patients have normal diploid cells in blood lymphocyte cultures and trisomy 21 cell lines from skin fibroblasts only.

As the risks to further children with Down's syndrome are greatly influenced by maternal age and by the chromosomal findings both in parents and children, special consideration must be given to each individual case and risks should be adjusted accordingly. The recurrence risk for regular trisomy 21 is low, $1-2\%$ unless there are predisposing factors (e.g., $+21$ parental mosaicism), which are known. There are some rare cases of direct transmission of Down's syndrome from mother to child. This does not necessarily imply that very low grade $1-2\%$ $+21$ mosaicism found in some laboratories in parents of children with trisomy 21 is of any value for the calculation of recurrence risks; prenatal diagnosis is indicated anyhow.

The maternal age effects are discussed elsewhere (Section 12). The recurrence risks quoted for Robertsonian t(14q21q) and t(21q22q) in a parent are usually $8-10\%$ if the mother is the carrier parent and $2-3\%$ if the father carries the translocation. The risk figures obtained from large-scale studies on amniocentesis material are probably more accurate but are subject to ascertainment bias as well as being uncorrected for late miscarriages that seem to affect Down's syndrome fetuses slightly more often than normal ones (see Section 12 on prenatal diagnosis). Where one of the parents is a carrier of a balanced reciprocal translocation involving chromosome 21, recurrence risks for Down's syndrome must be evaluated in the same way as for any other familial reciprocal translocation. Both history and cytogenetic findings should be taken into account.

10.2.3 *Trisomy 18 — Edwards' Syndrome*

Over 90% of these patients die within the first year. Female survival up to 18 years has been reported. There is usually a variety of congenital malformations, plus severe growth retardation (small-for-dates yet postmature). A face with a relatively broad and high forehead, a small jaw, and low-set malformed ears are characteristic. Typical flexion deformities of fingers, short sternum, narrow pelvis and limited hip abduction, short dorsiflexed big toes, rockerbottom feet, small genitalia including undescended testes (females reveal a prominent clitoris) and so on. Brain, heart, kidneys and gut are frequently malformed. The frontal falx is usually absent, and there is usually a ventricular septal defect. All are profoundly mentally retarded. Mosaic trisomy 18 is also found rather infrequently.

10.2.4 *Trisomy 13 — Patau's Syndrome*

Less than 10% survive beyond one year. Survival till 10 years has been reported. Many patients are blind and deaf and epileptic (not full fits), and all are profoundly mentally retarded. The phenotype varies from severe manifestation with cleft lip, micropthalmia and polydactyly to mild facial disproportion. Diagnosis is usually made on clinical grounds alone. Some characteristic features are microcephaly, sloping forehead, microphthalmia, arrhinencephaly, bilateral cleft lip and palate, malformed ears, hexadactyly, and so on. Congenital malformations involve the brain, eyes, heart and kidneys, and a variety of minor muscular and skeletal anomalies. Unbalanced Robertsonian translocations, most commonly t(13q14q), are sometimes found. The risk for balanced carriers of t(13q14q) of having translocation trisomy 13 offspring is probably less than 1%. Robertsonian t(13q14q) often causes male infertility. Mosaic trisomy 13 shows a less distinctive dysmorphic pattern but still rather severe mental retardation.

10.2.5 *Trisomy 8 Mosaicism*

Particularly surprising is the normal growth (height is usually *above* average for age) and the only moderate mental defect; sometimes even compatible with attendence at a normal school. Diagnostic features are a relatively large and square skull, broad nose, which is also prominent, everted lips and often small jaw, characteristic ear lobes, long thorax and slender body build, deep furrows in palms and soles, and hallux widely separated from the second toe. Other malformations are rare. Many adults have been described. The percentage of trisomic cells tends to decrease with age.

10.2.6 *Double Aneuploidy*

It seems likely that double aneuploidy is slightly more frequent than would have been expected by chance alone. Two autosomes are rarely involved for obvious reasons. Both events usually stem from the same parent.

10.2.7 *Triploidy*

The overwhelming majority of triploids abort spontaneously, usually in the first trimester of pregnancy. Those which survive to term are usually premature and die perinatally. 69,XXX or 69,XXY are the most common karyotypes. Patients with triploid mosaicism may survive into adult life with mild to severe mental retardation and other abnormalities. Somatic asymmetries are not uncommon.

10.2.8 *Tetraploidy and Diploid/Tetraploid Mosaicism*

These anomalies are usually not very amenable to cytogenetic investigation unless direct preparations are made. Tetraploidy in spontaneous abortions is associated with a missing fetus. Only two cases of tetraploidy in livebirths have been reported. Both died within a year, with serious defects.

10.3 **Autosomal Structural Rearrangements**

10.3.1 *General*

Structural anomalies of chromosomes all originate through chromosomal breakage and subsequent abnormal reunion. One or more breaks may occur in the same or different chromosomes. All have consequences at meiosis and might produce other abnormal karyotypes in the following generation. The prediction of their phenotypic effects must be based on the same general considerations as have been outlined in the notes above, which are also summarised at the end of Section 10. Investigation of the parents and possibly other members of the same family is always indicated. Chromosome breakage syndromes are discussed elsewhere (Section 11.2.4).

10.3.2 *Deletions*

Breaks can result in loss of parts of any chromosome. Deletions of large portions are usually incompatible with life. Some $10-15\%$ are due to a balanced rearrangement in one of the parents. These deletions represent partial monosomy/partial trisomy rather than true deficiency. The overwhelming majority $(85-90\%)$ are true deletions. The phenotypic effect can be predicted for those deletions that have been observed repeatedly. Schinzel: Catalogue of Unbalanced Chromosome Aberrations in Man (8) is a good source of information on this matter. For example, deletion of the distal part of chromosome 13 produces a well known syndrome with Greek profile, forward slanting incisors, absent or hypoplastic thumbs and bony syndactyly of fourth and fifth metacarpals. On the other hand, interstitial deletion of band 13q14 is often associated with retinoblastoma.

10.3.3 *Inversions*

Two breaks in the same chromosome may be followed by inversion of the intermediate segment before joining of the loose ends. At meiosis, the inversion gives rise to abnormal pairing involving an inversion loop (*Figure 2*). Cross-overs within this segment may give rise to gametes with duplications and deficiencies involving the two segments outside the inversion loop. The probability of a cross-over within the loop will obviously depend on its size and also on the general recombination frequency within that area. There is some evidence from other species that inversions tend to decrease recombination frequencies within the inverted segment. Inversions may also pair without a loop: (i) the inverted segment pairs normally but ends are unpaired, (ii) the ends pair normally but the inverted region pairs anomalously or not at all. Inversions vary in the extent to which they give rise to abnormal gametes. Some

inv(2)   cr.o. in BC

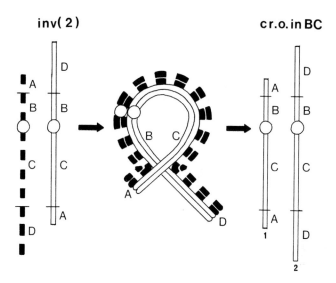

Figure 2. Pericentric inversion pairing configuration. Only the unbalanced duplication-deficiency gametes formed following crossing-over within the inversion loop are shown. The normal homologue of chromosome 2 is represented by dotted lines. All other chromatids are structurally abnormal in various ways as indicated by the lettering.

familial inversions such as, for example, the commonly observed small *pericentric inversion* of chromosome 2, do not seem to be of any serious consequence to the offspring, and must be regarded as normal variants, whereas others may cause reproductive failure or abnormal offspring. The general rule of interpretation of pericentric inversions is that the closer both breakpoints are to the telomeres, the greater the risk that unbalanced offspring will survive to birth. When the breakpoints are close to the centromere on a large chromosome there will usually be no risk of abnormal offspring, presumably due to the inevitable large deletions and duplications in the cross-over products. Predictions of reproductive effects of a particular inversion have to be based both on general considerations and studies of the particular family. Sometimes one can also use collective data from similar aberrations.

Relatively few *paracentric inversions* have been described. They differ from pericentric inversions in that they are more difficult to detect and cross-overs within the inverted segment should theoretically lead to only normal gametes or inviable gametes with dicentric isochromosomes or fragments, incompatible with fetal development. The risk should therefore be nil. This does not always seem to be the case; paracentric inversions are sometimes ascertained through a patient with congenital anomalies.

10.3.4 *Translocations*

Breakpoints on two separate chromosomes rejoin after reciprocal exchange of the distal segments.

(i) *Robertsonian translocation.* The commonest form of translocation is centric fusion involving two acrocentric chromosomes, the Robertsonian translocation. These may

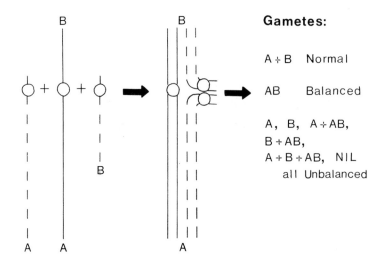

Figure 3. Pachytene pairing configuration and possible segregation in a carrier of a Robertsonian translocation. Non-disjunction in the second meiotic division could rarely give rise to further types of gametes. A translocation between chromosomes 14 and 21 is not uncommon. Only three types of gametes are known to produce viable offspring: (i) the normal ones (A+B in the figure), (ii) the balanced ones (AB in the figure) giving rise to translocation carriers, e.g., 45,XX,t(14q21q) and (iii) those containing both the t(14q21q) chromosome and a normal chromosome 21 (B+AB in the figure) producing translocation Down's syndrome, e.g., 46,XY,+21,t(14q21q) which could also correctly be written 46,XY,−14,+t(14q21q). This obscures the fact that there is an extra chromosome 21.

be monocentric or dicentric, which does not seem to matter very much. The corresponding small marker chromosome formed by the fusion of the satellited short arms of these chromosomes is usually lost but it can sometimes be seen as an accessory marker chromosome. Unbalanced gametes of Robertsonian translocations produce trisomy or monosomy for a complete chromosome. Only translocation mongolism (+21) and the much less frequent translocation Patau's syndrome (+13), are clinically important.

Figure 3 shows the segregation pattern at meiosis in a balanced carrier. If two homologous chromosomes are fused, only unbalanced products can be expected, unless the other gamete is nullisomic for the same chromosome. The vast majority of unbalanced 13/13 and 21/21 translocations are not inherited but arise *de novo*. Risks of abnormal offspring will be dealt with in the section on prenatal diagnosis (Section 12.3.2).

(ii) *Reciprocal translocation.* These translocations are also quite common. It is thought that about 1 in 1000 normal people carry a balanced reciprocal translocation which at meiosis can lead to the production of a variety of unbalanced duplication/deficiency gametes, together with the normal and balanced ones (*Figure 4*). For many reciprocal translocations, none or only one type of unbalanced segregation will lead to viable gametes. As always, the general rule for interpretation is that the size of the segments involved in the duplications and the deletions is critical. The number of bands involved has been used to estimate the risk of unbalanced progeny (see further in Section 12.3.2. on prenatal diagnosis). Gametes with large deletions cannot form viable off-

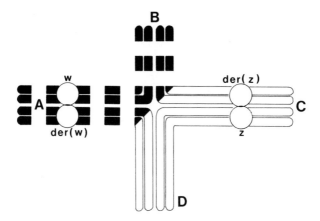

Figure 4. Pachytene pairing in the translocation heterozygote. Let w=AB, z=CD, der(w)=AD and der(z)=CB. Gametes transmitting w+z will form normal offspring, e.g., 46,XX, and those transmitting der(w)+der(z) balanced heterozygotes, e.g., 46,XY,t(w;z)(bkpt w;bkpt z). All other gametes are unbalanced. For example, a Y-bearing sperm with w+z+der(z) following 3:1 segregation would correspond to 47,XY+der(z),t(w;z)(bkpt w;bkpt z), Paris nomenclature, in the zygote. Note the paradoxical change in the meaning of the symbol t(w;z)(bkpt w;bkpt z) from its previous value of −w −z +der(w) +der(z) in the balanced heterozygote to become now a mere explanatory comment pointing at the der(z) symbol. Hence, 47,XY,t(w;z)(bkpt w;bkpt z),+der(z) is a very different karyotype. This ambiguity is frequently causing confusion. Furthermore, do not forget the consequences of crossing-over in the interstitial segments between the centromeres and points of exchange which will create a multitude of gametes that cannot arise by deficient segregation alone.

spring. Duplications are better tolerated. Many reciprocal translocations are familial, and have been inherited through many generations. Some cause high frequencies of abnormal offspring; others are relatively benign. Breakpoints are non-randomly distributed in familial balanced reciprocal translocations giving rise to abnormal offspring. However, most breakpoints are unique and fairly evenly distributed between the chromosomes. This implies that the same breakpoints are rarely found in two different families, but there are exceptions to this rule. The most frequently recurring type is t(11;22)(q23;q11), which produces 5−10% unbalanced offspring of female carriers. Only one type of unbalanced offspring has been observed, having an extra chromosome der(22),t(11;22) implying two partial trisomies, i.e., +11(q23→qter) and +(22)(pter→q11) following 3:1 segregation in the mother. Translocation carrier females have a high incidence of spontaneous abortions and perinatal deaths. The surviving affected children show growth retardation and mental deficiency, with a tendency for stereotypic movements and poor co-ordination, a characteristic face and, in the male, genital hypoplasia, i.e., micropenis, small scrotum and undescended testes. Because there are so many cases documented of this particular translocation, interpretation of the phenotype is quite easy, even if one should detect unexpectedly the unbalanced form *de novo* in prenatal diagnosis.

(iii) *De novo translocation.* Less than 10% of individuals with a *de novo* translocation show some features that might be caused by the rearrangement in spite of the fact that it looks perfectly balanced; perhaps the breakage and reunion event destroyed the function of the genes at the join. There is some evidence for this in autosome-

autosome translocations, but it can be particularly clearly demonstrated in females with *X-autosome translocations* involving band Xp21 and the X-linked recessive disease Duchenne muscular dystrophy − presumably they manifest the disease because the breakpoint on the X splits the Duchenne gene and their normal X is always inactivated. It is thought that inactivation occurs at random but is followed by selection against those cells where the der(X) translocation chromosome has been inactivated. Spreading of inactivation on to the autosomal part would make the cell functionally monosomic for the corresponding segment which presumably is lethal. About 50% of women with X-autosome translocations seem to be fertile: almost all affected men are sterile.

(iv) *Complex translocations.* Those translocations involving more than two chromosomes are sometimes seen and sometimes go undetected. Therefore, it is important to ensure that analysis is fully completed even if one rearrangement has already been found. Meiotic studies might be helpful in some cases, but are rarely done.

(v) *Non-disjunction in carriers of balanced translocations.* This seems to be more frequent than would be expected to occur by chance alone. The reason for this is unknown. The effect would cancel if current estimates of the frequency of translocation carriers in the general population were too low by at least a factor of two, which might seem rather unlikely (9).

10.3.5 *Insertions*

'Three-break rearrangements involving one or two chromosomes are referred to as insertions since they result from the excision of a segment following two breaks in one chromosome arm, and its insertion at a point of breakage in either the same arm or the opposite arm of the same chromosome, or in another chromosome. The order of the bands on the inserted segment in relation to the centromere at the new site may be the same as at the original site (direct insertion) or it may be reversed (inverted insertion)' [ISCN (1978) (1)].

During meiosis, two inversion loops must be formed to provide homologous pairing. Cross-over events within these loops create unbalanced products, some of which may look amazingly similar to the balanced parental rearrangement.

10.3.6 *Ring Chromosomes*

Two breakpoints on the same chromosome rejoin the interstitial segment forming a ring. The distal segments are lost. If a centromere is included in the ring, it may remain stable and divide properly in spite of the topological difficulties, but rings may also induce further changes, for instance, mosaicism or duplication/deficiency. Phenotypic expression of ring chromosomes is rather unpredictable.

10.3.7 *Terminal Rearrangements*

These represent end to end joining of telomeres. The resulting dicentric chromosome may remain stable through suppression of one of the centromeres. Very little is known about this interesting phenomenon.

10.3.8 *Small Accessory Marker Chromosomes*

The presence of an additional small chromosome of unknown origin is referred to as a marker chromosome. They comprise a mixed collection of structurally rearranged chromosome regions. About 90% of them are thought to be derived from the short arm and pericentric regions of the acrocentric chromosomes, and these are largely, but not exclusively, heterochromatic. Almost half of them seem to involve the short arm of chromosome 15.

10.4 **Clinical Findings in Sex Chromosome Aberrations**

The sex chromosomes behave differently from the autosomes. The usually mild manifestation of a sex chromosome anomaly contrasts sharply with the serious consequences of autosomal syndromes. The most characteristic feature of sex chromosome aberrations is in fact infertility, or reduced fertility, as is invariably seen in XXY males, in the absence of major malformations or mental retardation. Growth retardation is only typical for 45,X Turner's syndrome, which may also be associated with coarctation (narrowing) of the aorta and with other dysmorphic features. Increased height is seen in both XXY and XYY males. Mental retardation is either absent or mild, except in males with structural aberrations on their only X chromosome, and sex chromosome polysomy (48,XXXY or XXXX, 49,XXXXY or XXXXX, etc.), which profoundly affects the mental capacity. Translocations between sex chromosomes and autosomes may sometimes have serious consequences. The implications for the unbalanced offspring are always serious.

10.5 **Numerical Sex Chromosome Aberrations**

10.5.1 *Phenotypic Expression*

The phenotypic expression of numerical sex chromosome aberrations can more easily be understood if one remembers that the most important segments are presumably the non-inactivated parts of the X chromosome and the euchromatic part of the Y chromosome. These regions are partly homologous, and at least some genes located there behave as pseudoautosomal genes. The 45,X karyotype can therefore be thought of as a partial monosomy, and the 47,XXX and 47,XYY sex chromosome trisomies as partial trisomies for a very small segment. Several extra copies of the same segment will obviously cause worse effects. It may seem hard to reconcile the phenotypes of individuals having structurally abnormal sex chromosomes with such a simple model, but that is not necessarily true because X-inactivation is always abnormal when there are structurally abnormal X chromosomes.

10.5.2 *Sex Chromosome Monosomy 45,X Turner's Syndrome*

More than 90% of 45,X conceptuses die within the first trimester. Severely affected fetuses with cystic hygromas are frequently found at ultrasound examination in the second trimester. Newborns with 45,X often have lymphoedema of the back of the feet. Short stature, normal intelligence and primary amenorrhoea are the most important features later. More than half have malformations in the renal tract, and heart valve anomaly or coarctation of the aorta are also found occasionally. These physical abnormalities rarely need surgery or cause disease. The 45,X karyotype represents about half of the cases with this syndrome. At least one third of all patients with the Turner phenotype show chromosome mosaicism with one 45,X cell line. The other line frequently contains a normal 46,XX line or a structurally abnormal X in addition to the normal X and in general these patients show milder signs of Turner's syndrome than those with 45,X only.

10.5.3 *Sex Chromosome Trisomies*

(i) *47,XXX Triple X females.* These do not show a distinctive phenotype. There is sometimes a mild delay in mental development. Some may have reduced fertility. The offspring of these patients are said to be normal, but apparently there are no large-scale investigations to confirm this statement.

(ii) *47,XXY Klinefelter's syndrome.* The clinical features of Klinefelter's syndrome do not become apparent until after puberty, although advanced growth is already evident during childhood; there may be a mild delay in mental development. The prepubertal testes are normal but during adolescence they fail to develop. Gynaecomastia and other eunuchoid features may be present in a minority of cases. All 47,XXY males are thought to be sterile. Fertile ones are usually excused as having presumptive 46,XY/47,XXY mosaicism, even if the normal cell line cannot be found.

(iii) *47,XYY syndrome.* These males do not show a distinctive phenotype either, except that some have long legs and there is at least a five times increased risk of ending up in institutions for recurrent and violent delinquents (compared with 46,XY). There might be in some, mild mental retardation and a slightly reduced fertility, but nearly all lead quiet, undramatic and normal lives. Direct transmission of the extra Y from father to son does not seem to happen very frequently, but has been described.

10.5.4 *Sex Chromosome Polysomies*

Mental retardation is a common feature in individuals with more than three sex chromosomes. Tetrasomics are mildly to moderately mentally retarded, pentasomics moderately to severely retarded. They also show dysmorphic features and malformations to a variable extent, resembling autosomal syndromes.

10.6 **Structural Sex Chromosome Aberrations**

There is nearly always a bias of X-chromosome inactivation in XX females with structural sex chromosome aberrations. The normal X is inactivated in X-autosome translocations, and the abnormal X tends to be inactivated in XX females with structural aberrations of one X. It is not an all-or-none effect since inactivation is presumably initially random, but selection against unbalanced cell lines introduces the bias towards non-randomness.

46,XY males with balanced X-autosome translocations or Y-autosome translocations are usually infertile. Males with unbalanced X-rearrangements are usually severely mentally retarded. Y-rearrangements may be of no clinical significance, or give rise to Turner's syndrome.

46,XX females with balanced X-autosome translocations may be (a) normal, (b) have gonadal dysgenesis if breakpoints are within (X)(q13→q26), or (c) have multiple congenital anomalies and mental retardation.

Females with isochromosome Xq are more Turner-like than those with other X-chromosome rearrangments.

X-Y translocations may cause XX males. Insertion of an euchromatic segment of the Y chromosome into band Xp22 may cause male or fertile female phenotypes, presumably depending on the X-inactivation pattern.

10.7 Fragile X Syndrome

The Fragile X syndrome is an X-linked and common form of moderate to severe mental retardation, associated with a folate-sensitive fragile site in band Xq27.3. Almost all of the mentally retarded males in a particular family express the fragile site. Less than 1% of males that express the fra(X) have a normal IQ. Females carrying the fra(X) differ phenotypically with regard to its expression. The frequency of fra(X) positive cells in female carriers is directly correlated with mental impairment. Intellectually normal females express at a low frequency or not at all.

The clinical features are usually non-distinctive in younger boys, but their faces might be characteristically longer than normal, with prominent jaw and forehead and large ears. They may be autistic, hyperactive, or just mentally retarded. After puberty macroorchidism is a common feature; other physical signs are vague or entirely absent.

From the cytogeneticist's practical point of view it is important to score a large number of cells, not less than 100 and preferably 200, if a fragile X cannot be seen at a high frequency. Otherwise, for statistical reasons, one might easily miss it if the level of expression is of the order of $2-5\%$. This makes the scoring procedure rather time-consuming. It might prove necessary in busy laboratories to apply rather stringent criteria when accepting samples for fragile X investigation.

10.8 In Summary

The findings of a chromosome abnormality does not always imply multiple defects in morphogenesis, growth disturbances and mental retardation in the patient. Some anomalies cause no harm whatever. Their effects on the phenotype obviously depend on both the quality and the quantity of the genetic material involved. Some plan for investigating whether the chromosomal findings are relevant or not must be applied.

(i)   Is the karyotype balanced or not? Any extra genetic material as compared with the normal karyotype, or any missing material, should be recorded and evaluated in relation to the clinical findings. Check whether symptoms are similar to those of other cases described in the literature (8).

(ii)   If there is a genetic imbalance due to a structural aberration, missing genes are generally more deleterious than extra genes. The greater the imbalance the more serious, usually, the consequences. Some chromosome regions are apparently more important than others. A partial trisomy for one region might have far more serious consequences than partial trisomy for an equal-sized region on another chromosome.

The karyotypes of the parents might help to clarify the true nature of an unbalanced product of a structural rearrangement. For example, a deletion might be either a true deletion or a duplication-deletion product of a balanced rearrangement.

(iii) It should also be kept in mind that the breakpoints of apparently balanced structural aberrations may interfere with the function of the genes concerned by splitting them, or upset function of flanking genes due to position effects [e.g., females with X-autosome translocations and the X-linked recessive disease Duchenne muscular dystrophy − see also (iv)]. Such anomalies are often associated with unusual manifestations of the disease.

(iv) X-autosome translocations in the female usually upset the normally random inactivation pattern of the X chromosomes. That may have further implications for the phenotype.

(v) Carriers of balanced rearrangements may or may not have an increased risk of having unbalanced viable offspring. Data on this is sometimes available. They might also have an increased risk of producing children with seemingly unrelated trisomies such as trisomy 21 in the case of, for example, t(4;8). Prenatal diagnosis should be considered in all cases.

(vi) The majority of events which result in a whole extra chromosome, or one missing, are lethal. Only a few autosomal trisomic conditions are regularly found in liveborns and of these only trisomy 21 babies have a more than 10% chance of survival for at least a year. These trisomies are well described syndromes. Recurrence risks are usually low.

(vii) Numerical sex chromosome aberrations as well as most structural X anomalies in the female are generally much less deleterious than autosomal aberrations. The most characteristic feature is infertility.

(viii) Is the chromosome abnormality present in a mosaic condition? Interpretations of genotype-phenotype relations generally postulate that mosaic cases present a diluted or milder version of the corresponding non-mosaic condition. This might not always be true, but on the whole the higher the proportion of the aneuploid cell line the more severe is the clinical picture. Very little is known about the effect of aberrant cell lines in localised areas of the body although a few cases have been described in which the aneuploid cells seem to be present only in the malformed area.

(ix) Is there any chance that other family members might have, or be at risk for, children with the same anomaly? The family history is perhaps the most important piece of information we have for the interpretation of certain abnormal karyotypes, Karyotyping the parents is nearly always indicated.

(x) Does this case contribute any new information as regards our understanding of chromosomal disorders? If so, write it up and try to get it published!

11. ACQUIRED CHROMOSOME ABERRATIONS

Chromosome mutations in somatic cells show greater diversity than those that can be transmitted to later generations. The occasional cell with an acquired anomaly is usually of so little importance in the context of 10^{14} other cells in the human body that it can be ignored in the constitutional karyotype; but it may sometimes cause changes in its bearer. The commonest and most serious consequence is the development of cancer. Aneuploid clones of cells with specific chromosome rearrangements can be uniquely associated with distinct forms of neoplasia. Cytogenetic analysis of these cells may give invaluable information as regards the diagnosis, treatment and prognosis of the disease. Exposure to ionising radiation, radioactive chemicals, mutagens, viral infections and certain inherited pre-malignant disorders, may also lead to increased frequency of acquired chromosome aberrations. The cytogenetic features may be of diagnostic value. These anomalies are both time-consuming and difficult to analyse, but there is no way round these problems because their proper classification is a prerequisite for their interpretation.

11.1 **Nomenclature (ISCN 1978)**

The following terms are commonly used.

(i) A *chromatid-type aberration*, as opposed to chromosome-type aberration, involves a break of only one of the pair of sister chromatids of a metaphase chromosome. This may be a non-exchange aberration such as a *chromatid gap* (a non-staining lesion with minimal misalignment of the distal fragment) or a *chromatid break* (showing clear misalignment of the proximal and distal parts of the broken chromatid). *Chromatid exchange* aberrations within a chromosome may involve breaks at the same locus on both sister chromatids, thus creating a *SCE* (which can only be seen by autoradiography or BrdU staining techniques).

Alternatively, if an asymmetric rearrangement took place, a dicentric isochromosome and an acentric fragment may appear, provided all the loose ends are rejoined (*complete* exchange) and the breaks did not occur at the centromere. Exchanges between chromatids of different chromosomes may create *triradial, quadriradial* or *complex configurations*. Both homologues may be involved in the formation of a triradial or quadriradial configuration. *Incomplete exchanges* do not rejoin all loose ends that have been created at the breakpoints. Exchange aberrations may also be classified as *asymmetrical* or *symmetrical* depending on whether an acentric fragment is formed or not.

(ii) All chromatid-type aberrations can most easily be thought of as having occurred at or after DNA replication in the cell cycle. In the next generation cells they will appear as *chromosome-type aberrations* (involving both chromatids). The corresponding terms are *chromosome gaps, chromosome breaks, chromosome exchanges* (i.e., the constitutional aberrations described in Section 10). Some chromosome-type aberrations usually do not appear in constitutional karyotypes because they may be lost at cell division. For example, acentric fragments are subject to random loss at cell division. A special case of multiple small acentric fragments are the small *double minutes* often found in tumour cells and believed to represent amplification of genes. The same phenomenon in a more stable form is represented by *homogeneously staining regions (HSR)* as occasionally seen on rearranged chromosomes in tumour cells.

(iii) Total destruction of the genetic machinery can be seen as *chromosome pulverisation*. This may involve only one or a few chromosomes in a cell while the remaining chromosomes have a normal morphology.

(iv)   The presence of a multitude of aberrations in a single cell may create difficulties as regards the chromosome enumeration. The *chromosome count* should include all centric structures regardless of the number of centromeres in each structure. Acentric fragments are not included in the count. For example, the karyotype of a cell with a quadriradial configuration involving chromosomes 1 and 2, and an extra acentric fragment, could be recorded as 45,XY, − 1, − 2, + qr(2 cen), + ace [ISCN (1978) (1)].

(v)   *Clones of abnormal cells* can obviously arise from post-zygotic events creating a mosaicism of constitutional karyotypes. They may also represent tumour cell populations. Such clones are not necessarily homogeneous due to further changes that may have occurred during the different stages of tumour progression, or even during culturing the cells *in vitro*. The most frequent chromosome constitution of a clonal population is designated a *stem-line*. Other karyotypes are termed *side-lines* or *sub-lines*. In writing down the chromosome constitution of a tumour it is important to give information on the *date of observation*, whether it was a *direct preparation* or preparations from a cell population *cultured in vitro*, the number of cells counted and the number of each kind; the fully analysed cells should also be duly represented. The results will usually form a small table, from which some conclusion can be drawn. For example, the inference may be 'There is a stem-line of 46,XX,t(9;22) cells typical of chronic myeloid leukaemia with a Philadelphia (Ph^1) chromosome'. The variation between cells may be very great. Sometimes it is not even possible to establish a modal chromosome number but terms such as near diploid, hypodiploid and hyperdiploid, can always be used.

11.2 Chromosome Instability

The induction of cytogenetic variation is one of the most obvious effects of agents that can damage DNA and thus are biologically hazardous. Some of the physical mutagens, such as X-rays, interact with DNA and produce double-strand breaks at any stage of the cell cycle. Most chemical mutagens are S-phase dependent. Quite a few mutagens need to be converted to a biologically active form by metabolic processes. The cell itself can repair fairly extensive damage, correctly or incorrectly, through DNA repair mechanisms. Frequently, an important factor as regards the consequences of DNA lesions is how long the DNA exists in a damaged form before the DNA repair mechanisms cause a return to normal. Some individuals have recessively inherited defects in their repair mechanisms. They are more prone than others to certain forms of malignancy. For example, patients with the disease xeroderma pigmentosum, showing deficient repair of u.v. lesions, often get basal cell carcinomas, squamous cell carcinomas and malignant melanomas. However, it must be said that, excepting Fanconi's anaemia, evidence is weak for genetically impaired repair of DNA damage in the other two chromosome breakage syndromes ataxia telangiectasia and Bloom's syndrome. Other mechanisms may be involved.

11.2.1 *X-ray-induced Damage*

The primary effect of X-ray irradiation is the formation of ions; either atoms that are part of the DNA molecule itself or atoms in its immediate neighbourhood are ionised. This produces chemical reactions which may destroy the DNA locally. If cells are irradiated in G_1 the chromosomes react to the radiation as though they were single-stranded and a 'hit' will therefore correspond to a double-strand break in DNA. The broken ends can rejoin to form aberrant chromosomes that after replication in S will appear as chromosome-type aberrations. On the other hand, if irradiation occurs in S or G_2 then the already replicated chromatids are the units of breakage and reunion, which leads to chromatid-type aberrations. The number of induced chromosome aberrations increases linearly with the dose of radiation, although, as can be expected, the frequency of translocations and dicentric chromosomes increases with the square of the dosage. There is no threshold, $10-20$ rad causes detectable increases in abormalities, which means that cytogenetic analysis provides a reliable dosimeter for monitoring people who may have been irradiated, either inadvertently, or in the course of medical treatment. Chromosome abnormalities can be observed in lymphocytes many years after whole-body exposure to radiation. X-rays have also been shown to induce non-disjunction in developing germ cells, as well as tumours and gene mutations, of course.

11.2.2 *S-phase-dependent Chemical Mutagens*

Most chemical mutagens produce lesions in DNA that may be thought of as potential chromosome damage rather than actual breaks because they lead to chromatid-type aberrations only when the cell passes through the S-phase of the cell cycle. Lesions that are repaired before that will not produce any aberration. Unrepaired lesions may retain their ability to induce chromatid aberrations in any successive round of DNA replication. The formation of SCEs is much more sensitive to chemicals than is the formation of chromatid breaks and exchange aberrations. SCE, which is thought to occur at the point of replication, is therefore the method of choice to study the *in vitro* effects of most chemical mutagens. Ionising radiation, on the other hand, is a rather inefficient inducer of SCEs.

11.2.3 *Virus Infections*

Although virus infections may produce chromosome aberrations as well as malignancy, this is a rather uncharted subject that has more unknown elements than real data.

11.2.4 *Inherited Predisposition to Spontaneous Chromosome Breakage and Malignancy*

(i) *Bloom's syndrome* is caused by an autosomal recessive gene. It is an extremely rare condition characterised by small stature, but an otherwise well-proportioned body build, and severe disturbance of immunity. There is a much increased risk of leukaemia and other forms of cancer. The cytogenetic findings are striking. One to 15% of PHA-stimulated lymphocytes show homologous chromatid-type quadriradial interchange configurations. Spontaneous SCE frequencies are many times higher than normal; which so far has not been found in any other condition.

(ii) *Fanconi's anaemia* is a rare autosomal recessive disorder of the bone marrow causing pancytopenia, usually between the ages of 4 and 12 years. Other clinical features include thumb and radius anomalies, growth retardation and skin pigmentation. There is an increased risk of acute leukaemia, but relatively few other malignancies. The chromosomal changes include chromatid and chromosome gaps, fragments, translocations, dicentrics, rings, inversions and endoreduplicated chromosomes in a high frequency in lymphocytes as well as in skin fibroblasts. The breaks are supposedly random. These patients do not show the specific homologous interchange aberrations seen in Bloom's syndrome. Cells from Fanconi patients are extremely sensitive to cross-linking mutagenic chemicals such as mitomycin C although SCE frequencies do not increase above normal.

(iii) *Ataxia telangiectasia* is also a rare syndrome with autosomal recessive inheritance. Cerebellar ataxia is usually observed when the child begins to walk, and is progressive. Other neurological features such as dysarthria and oculomotor problems also occur. Telangiectasia may appear on the bulbar conjunctiva and some areas of the skin at a later stage. Respiratory tract infection may be a problem in some patients, but not all, and lymphoma or leukeamia may occur in about ten per cent of patients. All A-T patients have a defect in cellular immunity and a proportion also have a defect in humoral immunity. Gene carriers have been suggested to have a higher rate of developing some malignancies compared with normal individuals. The blood lymphocytes may not respond very well to PHA. There is a slight increase in cells with dicentrics, rings, gaps and breaks. More striking however is the increased level of translocations or inversions involving mainly chromosomes 7 and 14. Clones with such rearrangements may emerge; those with t(14;14) translocations proliferate more readily than others. Lymphocytic leukaemia with chromosome rearrangements involving 14q11−12 and 14q32 have been observed. Cells in culture from these patients are unusually sensitive to the induction of chromosome damage by X-rays. SCE frequencies are normal.

11.3 Tumour Cell Populations

Most tumour cell populations show chromosome abnormalities which, to a variable extent, may reflect both tissue of origin and the transforming agent. Superimposed on the specific changes is an often even greater random variation indicating malfunctioning of the spindle, chromosome breakage, delayed or deficient DNA replication and a general escape from the stringent control that governs proliferation of normal diploid cells *in vivo*. One can assume that all these chromosomal errors as such do not reflect a positive advantage of the tumour cells, but merely a deficiency of some kind which may be a prerequisite for the malignant transformation.

Let me briefly discuss a few remarkable steps of progress that have recently been made in cancer research. A variety of chromosomal changes including endoreduplicated chromosomes, structural rearrangements, double minutes and homogeneously staining regions, may result from over-replication of DNA within a single cell cycle. Molecular DNA techniques have provided evidence that cellular oncogenes may be both activated and amplified by such processes. The normal role of most of these genes seems to be in growth regulation. Presumably they define various elements of 'mitogenic pathways'.

Like any other gene, their proper functioning can also be destroyed by mutation or translocation. For example, Burkitt's lymphoma cells induced by Epstein-Barr virus show a specific translocation between 8q24 (containing the c-*myc* oncogene) and any of the chromosome segments 14q32, 2p11, or 22q11 (containing immunoglobulin genes). Bringing the immunoglobulin and the c-*myc* oncogene into the proximity of each other in the differentiated and already transformed B lymphocyte results in high levels of c-*myc* expression which, in turn, may provide a further proliferative stimulus for the affected cell.

Growth stimulation is usually not enough to make a normal diploid cell become malignant. The transfection with cellular oncogenes isolated by molecular cloning techniques does not result in cells that are competent to form tumours *in vivo*. Current theory postulates that malignant change occurs by a process involving at least two steps.

Such theory is partly based on the fact that predisposition for certain tumours may be inherited as Mendelian traits. For example, retinoblastoma is a malignant tumour of the eye in young children. The majority of these cases are sporadic but some are caused by an autosomal dominant gene located in band 13q14. In a proportion of the sporadic cases tumour formation is associated with interstitial deletion of this band in the constitutive karyotype. Molecular genetic analyses have shown that homozygosity for this region can almost invariably be demonstrated in the tumour cells even when the patients do not show a deletion on cytogenetic examination. Apparently, both mutation and somatic recombination or other chromosomal events must have occurred during tumour induction, making a recessive lesion appear as a dominant trait because it presents in the homozygous form. The predisposing germline mutation is also found in the homozygous form in tumours from children having the inherited propensity; in spite of the fact that all other cells in their body are heterozygous.

Another line of evidence is based on the suppression of tumorigenicity in hybrids of normal and oncogene-transformed cells. Tumour-forming ability is inhibited in the hybrid cells, even in the presence of an activated oncogene.

In summary, one may conclude that although most tumour cell populations are of clonal origin and possess cytogenetic characteristics which are not present in any other cells of the individual, only a few of these changes appear to be causally related

Table 3. Maternal Age-specific Rates for Trisomy 21 Detected at Amniocentesis.

Maternal age at amniocentesis	Number of pregnancies	Frequency of trisomy 21	
		Fraction	(%)
35	5409	1/285	0.35
36	6103	1/174	0.57
37	6956	1/148	0.68
38	7926	1/124	0.81
39	7682	1/91	1.09
40	7174	1/82	1.23
41	4763	1/68	1.47
42	3156	1/46	2.19
43	1912	1/31	3.24
44	1015	1/34	2.96
45	508	1/22	4.53
46	232	1/12	8.19

Data from European Collaborative Study (10).

119

to the development of the tumour. Much remains to be explained, even if the pattern that emerges is that specific lesions in the genome seem to be needed to produce a particular type of tumour. Cytogenetic analysis may reveal a few of these lesions and is frequently of diagnostic value.

12. PRENATAL DIAGNOSIS

12.1 Screening for Down's Syndrome

A collaborative European study on 52 965 amniocenteses in women 35 years or over (10) provided the necessary data for rate of detection of chromosome anomaly at amniocentesis. *Table 1* shows a comparision of rates of chromosomal abnormality in pregnancies from women of age 35 years or over, with rates determined from surveys of the newborn of all maternal ages. *Table 3* shows maternal age-specific rates for trisomy 21. By necessity, the detection rate following first trimester chorionic villus sampling will be very much higher: accurate statistics are not yet available.

It has recently been shown that Down's syndrome fetuses are associated with a lower level of maternal serum AFP as well as a lower level of amniotic fluid AFP than normal. This could be used in combination with the maternal age factor to improve the prediction as to which pregnancies are at risk, especially in the 30−35 years of age group of women (6).

12.2 Patients' Information

An example of our explanatory leaflet given to the patients before amniocentesis is shown in *Table 4.*

Table 4. The Detection of Birth Defects by Amniocentesis.

You have been referred to the Maternity Outpatients, John Radcliffe Hospital. There will be ample time for discussion and asking questions at the clinic, but it may help you to read this short description of some of the advantages and disadvantages of the test beforehand, so that you can think things over before you see us. Your husband should come with you if possible.

What is amniocentesis?
Amniocentesis is the taking of a small sample of fluid from the 'sac' around the baby. The test is ideally done between the 16th and 18th week of pregnancy, but if there is too little fluid the test may have to be delayed for 1 or 2 weeks.

Why is it done?
The cells which have been shed from the baby contain chromosomes; these can be examined for evidence of Down's syndrome (mongolism), and other rarer chromosome disorders. An increased level of alpha fetoprotein (AFP) is usually present in the amniotic fluid if the baby has spina bifida; this protein is always measured.

Who needs the test?
Amniocentesis is only able to detect some abnormalities. Even if the mother has already had an abnormal baby, the test may not be helpful. The following are the main indications.
1.   *Maternal age*: for a healthy couple with no family history of Down's syndrome the risk of giving

birth to a baby with this condition is related to the age of the mother, approximately as follows:

Woman's age 25   means risk of Down's (mongolism) is 1 in 1200
Woman's age 30   means risk of Down's (mongolism) is 1 in 900
Woman's age 35   means risk of Down's (mongolism) is 1 in 400
Woman's age 37½ means risk of Down's (mongolism) is 1 in 200
Woman's age 40   means risk of Down's (mongolism) is 1 in 100
Woman's age 43   means risk of Down's (mongolism) is 1 in   50

2.   *Relative with Down's syndrome:* the first step would be to do a blood test on you, your husband or the affected person. From this we could find if amniocentesis would be helpful; usually it is not.
3.   *Previous Down's syndrome:* if you have already had a child with Down's syndrome the risk of having a second child affected is about 1 in 100 in addition to the risk above.
4.   *Previous spina bifida:* if you have had a baby with spina bifida or anencephaly the risk of this happening again is about 1 in 30.
5.   *High serum AFP:* sometimes an unusual test result in pregnancy (spina bifida blood test taken at 16 weeks) requires further investigation in Oxford and amniocentesis may be one of the tests recommended. Less than 1 in 10 of these mothers are found to have anything wrong.

Is any preparation required?

Please bring documentation of your blood group if you are from outside Oxford. A normal breakfast or lunch may be taken. Ultrasound scanning is often easier if you have a full bladder. You should not drive yourself immediately after the procedure.

How is it done?

An ultrasound scan is performed to examine the baby and define the fluid present. A small area of skin on the abdomen is made numb with a local anaesthetic. The amniocentesis needle is inserted through the skin into the womb and a sample of fluid withdrawn. This is not usually painful, but a feeling of discomfort for up to an hour or so after is common. A blood sample is taken from your arm before the test.

What are the risks?

The ultrasound scan is harmless to both mother and baby. It is extremely unlikely that the baby will be directly injured by the needle. However, insertion of the needle may provoke a miscarriage, even when the baby is perfectly normal. It is difficult to give accurate figures for this risk, as women who have not had an amniocentesis may miscarry at this time of pregnancy; the added risk is less than 1 in 100 per test. If the test does bring about a miscarriage it will probably do so within 10 days. The risks are higher with twins.

Are the risks worth taking?

Down's syndrome and other chromosomal abnormalities can almost always be detected, as can spina bifida. Spina bifida is usually seen on scanning, but a good view is not always possible.

What should you do after the test

You should remain at the hospital for an hour after the test. For the rest of the day you should take things easy, and not drive. Slight bruising of the skin may sometimes appear where the needle was inserted. If you have any vaginal bleeding or spotting this need not mean a miscarriage, but you should rest in bed and phone your GP.

How soon are the results known?

Normal spina bifida test results: it takes up to a week for the AFP tests to be completed. If the amniocentesis was performed because of a high risk of spina bifida then normal results are sent to you and your doctor by post. If the results have not been received within 10 days of the test, then telephone the AFP Screening Laboratory (Oxford 0865/817489).

Table 4. continued

Normal chromosome results: your chromosome test takes up to 4 weeks to complete. Normal results will be posted to you with copies to the doctors concerned. You will only be informed of the sex of your child if you request it. The spina bifida results will be included with the chromosome results if amniocentesis was done for maternal age, or other chromosome reasons. Please phone exactly 1 month after the test if you have not heard (Oxford 0865/62834).

Only normal results will be sent by post. In other circumstances we will communicate through the doctors.

Failure to obtain a result

Very occasionally the baby's cells fail to grow. This is usually known within 10 – 14 days of the test. This does not mean that there is anything wrong with the baby. You will be informed immediately and, if the indication was for maternal age, invited back. For some people, e.g., those having amniocentesis to exclude spina bifida, there is usually no indication for the test to be repeated.

Change of address

Please notify us at the Department of Medical Genetics if you move house before the baby is born, and include the name and address of your new GP.

12.3 Interpretation of Prenatally Diagnosed Abnormal Karyotypes

12.3.1 *Fortuitous Diagnoses*

(i) *Numerical sex chromosome aberrations.* These are frequently encountered in the course of prenatal screening for Down's syndrome. Not everyone would agree that 45,X Turner's syndrome, 47,XXX triple-X female, 47,XXY Klinefelter's syndrome and 47,XYY male, are sufficient grounds for termination of pregnancy. Experience shows that some patients will choose the option of termination of pregnancy while others will not. It seems important that parents, after proper counselling, are allowed to reach a decision themselves, without imposition of personal views of laboratory or clinical staff on this matter.

(ii) *Balanced structural rearrangements.* These are also seen frequently. Since very few parents have had routine chromosome analysis before they conceive, such parental analyses will have to be carried out urgently in order to gain more information about the particular anomaly. Most rearrangements are inherited aberrations, which is reassuring. Only a minority will have arisen *de novo*, or turn out to be more complicated than was originally thought. This procedure may sometimes lead to parental anxiety so some cytogeneticists feel reluctant to ask for parental bloods, and would rather wait until the baby is born. However, it is very hard to defend not having done a full investigation. Withholding information on the fetal karyotype from the parents on the grounds that this information is not good for them may be well meant but it is hardly a rational decision. It might be difficult to establish long-term contacts with the family later. If parents are not informed while the case is under review no-one will remember the translocation the next time the patient becomes pregnant. Consequently the mother, who might well opt for not having an amniocentesis next time, may give birth to a malformed baby with an unbalanced form of the translocation. Such problems might also appear in children of close relatives (undetected carriers).

Table 5. Amniocentesis Outcome for Known Carriers of Structural Rearrangements.

Type of rearrangement	Carrier parent	Number of pregnancies[a]	Results					
			Normal		Balanced		Unbalanced	
			No.	(%)	No.	(%)	No.	(%)
Robertsonian t(13q14q)	Mother	157	69	(44)	88	(56)	0	–
	Father	73	27	(37)	46	(63)	0	–
t(14q21q)	Mother	137	48	(35)	68	(50)	21	(15)
	Father	51	20	(39)	31	(61)	0	–
t(21q22q)	Mother	20	11	(55)	6	(30)	3	(15)
	Father	2	0	–	2	–	0	–
Reciprocal translocation	Mother	378	168	(44)	166	(44)	44	(12)
	Father	231	97	(42)	107	(46)	27	(12)
Inversion (except 9)	Mother	67	32	(48)	30	(45)	5	(8)
	Father	51	14	(27)	35	(69)	2	(4)

Data from the European Collaborative Study (11).
[a]Note the bias of ascertainment. Reduced fertility in male carriers of Robertsonian translocations is well known; and also the mother would not come for amniocentesis unless she was pregnant.

12.3.2 *Inherited Structural Rearrangements: Phenotypic Effects and Recurrence Risks*

The risk of an unbalanced offspring to a known carrier of a balanced chromosome rearrangement depends on the type of aberration, the chromosome involved, the location of breakpoints, and also the sex, age and mode of ascertainment of the carrier. The basic considerations have already been discussed in previous sections. The mode of ascertainment can be used as a rough indicator of risks in the absence of specific data. If the chromosome aberration was found during investigation for recurrent miscarriage the risk would be low, perhaps less than 5%. On the other hand, if karyotyping was instigated because of the birth of a handicapped child, the risk for recurrence would be of the order of 20%. The overall risk for a carrier of a translocation karyotype is quoted at about 10%. Rates for specific anomalies, for use in counselling, were also obtained by the European Collaborative Study (11) (*Table 5*).

(i) The Robertsonian t(13q14q) is the commonest form of this type of anomaly and is fortunately associated with a very low risk, less than 1% of translocation trisomy 13 in the offspring. Robertsonian translocations involving chromosome 21, for example t(14q21q),t(21q22q), and so on, imply risks of the order of 15% when the mother is the carrier parent. No affected fetus was found among the 52 pregnancies where the father was the carrier. Selection against the unbalanced sperm may have a role in this phenomenon. In the offspring there is an excess of balanced carriers compared with those with a normal karyotype.

(ii) Reciprocal translocations give an overall rate of 11.7% unbalanced offspring at amniocentesis. The rate for carrier mothers did not differ from the rate in carrier fathers. Balanced carriers were as frequent as those with normal chromosomes among the offspring. If the duplication-deletion segment extended to between one and two chromosome bands, on the 300 band per haploid set level, the proportion of unbalanced progeny was 33.7%, an imbalance of three to four chromosome bands gave unbalanced progeny in 17.5%, and an imbalance of five to six chromosome bands gave unbalanced progeny in 8.8% of the cases. Interestingly, the distribution of the chromosomes involved in reciprocal translocations was different when ascertainment was through spontaneous abortions, for which a random distribution of breakpoints according to the relative size of the chromosome could be assumed, as compared with ascertainment through an infant with unbalanced anomaly, for which a clearly non-random distribution with over-representation of chromosomes 4, 9, 11, 18, 21 and 22 could be shown. The notorious t(11;22) was observed in several cases. All the others seemed to have unique breakpoints, which makes the evaluation of specific risks difficult.

(iii) Pericentric inversions, excluding the common inversion of chromosome 9 in which cross-over within the inverted segment has never been observed, gave a proportion of 5.9% unbalanced progency at amniocentesis. Consideration should be given to the actual location of the breakpoints. Small pericentric inversions are usually associated with a very low risk indeed (see further Section 10.3.3).

12.3.3 *De novo Structural Rearrangements*

These fall into three main categories:

(i)   unbalanced rearrangements;
(ii)   apparently balanced rearrangements;
(iii) small supernumerary marker chromosomes.

(i)   The first category usually represents no ambiguity of interpretation of the phenotype.

(ii)   The second category, *de novo* balanced rearrangements, on the other hand, present serious problems of interpretation. Since there is no other family member from which one can judge whether the rearrangement is truly balanced or not, one has to resort to collected data which indicates that the risk of an abnormal phenotype is about 8% (12). This risk figure has 95% confidence limits between 1% and 14%, which means that available data are not sufficient to rule out a risk of abnormality which is no higher than the usual rate of anomalies at birth (1−2%). Consequently, some authors claim that there seems to be no justification for recommending termination in cases where structural rearrangements are detected prenatally unless there is unequivocal evidence of chromosomal imbalance. According to my own view, there is no justification, after giving proper risk assessment, for recommending termination in any case of chromosomal abnormality. The decision should be left to the parents.

(iii) Non-familial supernumerary, or accessory, marker chromosomes detected *de novo* at amniocentesis also produce counselling difficulties since they may originate by a variety of different mechanisms and from various chromosomes, and quite frequently they are present in the mosaic state. Current estimates range from 5% to 30% risk of an abnormal phenotype. Satellited marker chromosomes might be more benign than non-satellited, although this has not been proved conclusively. Familial ones are nearly always innocuous.

12.3.4 *In Summary*

The interpretation of prenatally diagnosed abnormal karyotypes is usually straightforward. However, the detection of a *de novo* balanced rearrangement or a supernumerary marker chromosome presents problems requiring speediness in further laboratory investigations, and also problems of interpretation, since unreliable risk figures and rather an ambiguous description of the possible phenotypic effects will have to be conveyed to the parents. It is important that parents themselves are given the chance to decide whether to continue the pregnancy or to have a termination.

Large pericentric inversions with the breakpoints near the telomeres are especially treacherous and so are those reciprocal translocations that produce the smallest amount of detectable deletion at meiotic segregation (i.e., those that have one breakpoint near a telomere). Exchanges between two telomeres are very difficult to detect. In the case of inherited structural rearrangements in known carriers, these various factors should be considered before amniocentesis or chorionic villus sampling is actually performed.

12.4 Interpretation of Chromosome Mosaicism and Pseudomosaicism in Amniotic Fluid Cell Cultures

12.4.1 *A Note on Recommended Practice*

Chromosome mosaicism in amniotic fluid cell cultures cannot be studied in a meaningful way unless the sample is divided between three or more culture vessels. This approach provides a series of parallel, but individual cultures which have been established independently from one another.

Mosaicism found in one or more of these cultures may be interpreted differently depending on the way in which the cells have been harvested. The 'suspension harvest' techniques described in Chapters 1 and 5, which are normally applied to flask and Leighton tube cultures, disrupt the individual colonies and provide a series of essentially identical slides from a cell suspension in which a certain proportion of cells may show a particular anomaly. The '*in situ*' harvest techniques, which are used for coverslip and chamber cultures, allow colonies to be preserved intact so that each metaphase can be studied in its original context. This may allow easy differentiation between true mosaicism and pseudomosaicism which has arisen during cell culture *in vitro*. For example, a few abnormal cells in an otherwise normal colony is nearly always indicative of a cultural artefact.

Our current practice for '*in situ*' preparations is to analyse, where possible, at least one cell from each of ten different colonies distributed over four coverslips. If an abnormality is found the particular colony is examined further. In more than 95% of these cases the abnormality is restricted to a single cell or a small sector of the colony and can thus be ignored as a cultural artefact. If the colony should turn out to be homogeneously abnormal the further analysis will depend on the particular anomaly found (see Sections 12.4.2 and 12.4.3).

'Suspension harvest' preparations require a slightly different approach. The US survey on chromosome mosaicism and pseudomosaicism in prenatal diagnosis (13) recommended that at least ten cells should be analysed from each of two culture vessels.

If an abnormality other than a missing chromosome is found in one of these analyses, a further ten cells should be examined from the normal culture, and 20 cells should be examined from the third vessel. Single cell hypodiploidy is ignored here. There is usually no point in examining further cells from a vessel where an abnormal clone has already been found.

Whether or not there is any advantage in studying original cell colonies rather than homogeneous preparations is perhaps debatable. Certainly reseeding or migration of both normal and abnormal cells may occasionally obscure the picture on a coverslip. Many laboratories, including ours, do, however, prefer to use this approach.

12.4.2 *Pseudomosaicism Including Single Abnormal Cells*

Current practice detects *single abnormal cells* (excluding single cell hypodiploidy) in about 3% of amniotic fluid samples. The frequency varies with the culturing techniques used and also with the quality of the spreads. Monosomy may simply result as a preparation artefact during processing, and so may trisomy, although this is less likely. Some workers feel that most of the trisomic cells found in amniotic fluids represent extra-embryonic cells of unknown origin. The evidence for this would be the observation that trisomic colonies on coverslip cultures left undisturbed are usually homogeneous and not variegated, whereas colonies containing structural aberrations are usually mixed colonies which also contain normal cells. Most structural abnormalities should therefore presumably have arisen during *in vitro* culturing. Chromosome breakage is frequently a result of poor growth conditions in toxic media, or viral infection. Based on experience, it is common practice to ignore single cell abnormalities if the procedure outlined in the note of recommendations above has been followed. Obviously this does not exclude the possibility of a true fetal mosaicism but, according to statistics (14), all hopes of detecting and confirming a low grade mosaicism in amniotic fluid cell cultures can only be based on expectations of defeat. Given limited resources, it is simply impractical to score a sufficient number of cells distributed between a sufficient number of flasks or colonies. However, only exceedingly rarely would single cell abnormalities be expected to indicate true fetal mosaicism in conjunction with a corresponding birth defect; odds are certainly less than 1 in 100. For those cytogeneticists who wish to play safe the type of abnormality observed might decide whether extra flasks should be examined, for example, trisomy 21 is always cause for concern, whereas trisomy 10 is usually dismissed without much further hesitation because it has apparently never been found in a trisomic condition in a liveborn infant. Some chromosomes are typically involved more often than others in pseudomosaicism. Trisomy 2 and trisomy 20 are especially common, but trisomy 7, X, 9, 17 and 21 also seem to be more frequent than would have been expected on a random basis. The reason for this can only be speculated upon. The tendency is particularly clear in *multiple cell pseudomosaicism* [i.e., a clone of at least two abnormal cells confined to a single or multiple colony on one coverslip or, alternatively, a single flask (*Figure 5*)]. Similarly, monosomy for the smaller autosomes, 16−22, and the sex chromosomes, X and Y, are more common than others in multiple cell pseudomosaicism. Structural abnormalities occur approximately in proportion to the size of the chromosomes involved.

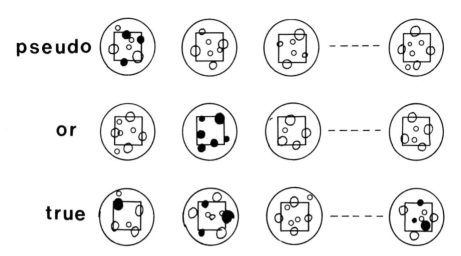

pseudo

or

true

Figure 5. Multiple cell psuedomosaicism versus true mosaicism. Modified from (17). Three rows of coverslip cultures of amniotic fluid cells in Petri dishes (which could also represent tissue culture flasks) showing the difference between multiple cell pseudomosaicism and true mosaicism as defined in the text. The first row of four, or more, cultures from one patient has one coverslip with three different colonies of the same kind of chromosomally abnormal cells (black dots) which are not found in any other culture vessel having normal colonies only (white dots). This would be classified as pseudomosaicism. The second row is a sample from another patient. One coverslip shows abnormal cells exclusively. All other colonies are normal. This would also be classified as pseudomosaicism (although it seems more likely that it represents a mix-up of samples). The third row represents true mosaicism. Note that one Petri dish has normal cells only. N.B., the definition is very crude and more satisfactory in practice than in theory.

These aberrant cells do not tell us anything about the fetus. Amniotic fluid cells grow clonally from a few of the original desquamated fetal cells. Non-disjunction and structural rearrangements frequently occur in cultured cells. It is therefore not surprising that a clone of multiple abnormal cells found in a single colony or in a single flask is usually without clinical significance. Out of 206 cases with pseudomosaicism involving multiple cells in the US survey on chromosome mosaicism (13) only two cases would have been diagnosed as a true mosaic from the follow-up cytogenetic studies. The experiences from other studies are very similar.

The incidence of pseudomosaicism involving multiple cells, or one or multiple colonies, in only one culture vessel, is of the order of 1% with current techniques. Once diagnosed no further action is usually indicated. It is certainly not an indication for therapeutic abortion unless further studies confirm a true fetal mosaicism. In the exceptional case, repeat amniocentesis or fetal blood sampling might be performed, especially in cases where the number of cells and flasks, or coverslips, available for analysis is insufficient to rule out a medium to high grade true mosaicism or, alternatively the suspicion that maternal cell contamination might be the source of the normal cells.

12.4.3 *True Mosaicism*

The presence of an abnormal cell line involving multiple cells distributed over more than one culture vessel should be regarded as a strong indication of true fetal mosaicism. In the majority of these cases an abnormal karyotype can be confirmed by chromosome analysis of the fetus or infant. The incidence of true mosaicism in amniotic fluid cell cultures is usually of the order of 0.25%. About one quarter of these fetuses will have an abnormal phenotype.

(i)   The phenotypic manifestations of numerical sex chromosome mosaicism, which amounts to about one third of the cases, are usually absent or mild, as can be expected from what is known about adult subjects with these types of mosaicism. About 10% are phenotypically abnormal at birth.

(ii)   Autosomal mosaicism excluding structural abnormalities contributes another one third of the cases: 30−40% of these show an abnormal phenotype. Trisomy 20 and 21 are the most common findings. The risk of having an abnormal phenotype is obviously higher if chromosomes 21, 18, 13, 8, 9, 22 and 20 are involved as compared with other autosomes, e.g., 2, 15, 10, etc., which usually cannot be confirmed upon follow-up cytogenetic studies.

In the case of prenatally diagnosed trisomy 20 mosaicism which presents a serious problem, about one sixth of the fetuses have been grossly abnormal, and five sixths of fetuses/liveborn infants have shown a normal phenotype and normal cells only, given the limited follow-up cytogenetic studies.

(iii) Structural abnormalities, including marker chromosomes and acentric fragments, which contribute the remaining one third of cases, can usually be confirmed on follow-up. The most unbalanced cell lines have a tendency to disappear altogether. Predictions of a normal phenotype can sometimes be made such as, for example, in the case of a parentally transmitted balanced reciprocal translocation. Information concerning the phenotypic manifestations of a particular marker chromosome inherited in mosaic form through several generations is also sometimes available. The overall frequency of abnormal phenotypes in this group is about 25%.

(iv)   Diploid/triploid mosaicism is rare ($< 1/100$) but presumably always associated with an abnormal phenotype.

12.4.4 *Contamination with Maternal Cells*

About 0.2% of amniotic fluid cell cultures show 'mosaicism' of the 46,XX/46,XY type. This is usually due to maternal cell contamination. It seems to be generally accepted that no further action has to be taken unless follow-up studies show ambiguous genitalia or that the wrong sex assignment has been made.

12.4.5 *In Summary*

Current practice of cytogenetic analysis of amniotic fluid cell cultures will identify medium to high grade true fetal mosaicism which may or may not be an indication for therapeutic abortion, depending on the cytogenetic anomaly in the individual case. Pseudomosaicism should either be dismissed or confirmed as true mosaicism by further studies.

13. CONSIDERATIONS OF REPORTING

13.1 **Ascertainment**

Before reporting a chromosome result one ought to appraise the analysis results and their interpretation in relation to the reason for referral. Intuitively, we all have a vague idea of the likelihood of finding a chromosome aberration in a particular case. Ascertainment through a previous child with a chromosome rearrangement will in most cases restrict the investigation to finding out whether a parent is a carrier of the same rearrangement and, if so, if there is anybody else that ought to be investigated.

An obligatory check is whether the chromosomal sex agrees with the phenotypic sex. Most cases of the wrong sex assignment are clerical errors. Some cases represent mix-up, either from errors in labelling the sample when it was taken, or because the laboratory committed a fault. Very rarely, the sample has unknowingly been taken from a genuine case of testicular feminisation, or from a 46,XX male. Fortuitous findings of abnormal karyotypes must be considered carefully before they are reported. Inevitably the clinician will ask for some plan of action on how to convey this to the patient without causing unnecessary anxiety or harmful effects. Most likely he would also welcome advice on further investigations.

13.2 Urgency

Ideally, all samples should be analysed the same day preparations are made. Due to pressure of work and other factors this is rarely possible except in cases with a high priority. Most laboratories use guidelines for giving priority to the individual case. Our current system classifies fetoscopic bloods and bloods from newborn babies as urgent. These can usually be reported within $3-4$ days of receipt of the sample. Other blood samples might also be classified as urgent, depending on the circumstances.

Amniotic fluids and chorionic villus samples for the purpose of fetal chromosome analysis are also high priority cases, which should be reported as soon as possible. It is not always possible to keep pace with a torrent of incoming specimens. Under such circumstances one can always screen the samples by counting a few cells, select the abnormals for immediate processing, and let the final analysis of the supposedly normal ones wait for a couple of days until the peak is over. The chorionic villus results are usually given as a preliminary report on direct preparations, followed by a formal report a couple of weeks later on cultured cells.

13.3 Confidentiality

The reporting system must be designed to avoid illegitimate breaks of confidentiality. It is good practice not to give out reports over the telephone unless you know for sure to whom you are speaking. Telephone messages are sometimes necessary and should in all cases be followed up with a written confirmatory note sent out on the same day with what was actually said during the conversation. A copy of the letter should be kept for filing. The patient, especially in prenatal diagnostic cases, sometimes does not even want her own general practitioner to know the results, and one must respect that wish. It is good practice to do what one is told in such circumstances. The normal report should be in writing to the consultant who instigated the chromosome investigations. A copy should be kept for filing purposes.

13.4 Format

The report should contain information on the patient's name, and address and date of birth for identification purposes, the tissue investigated and the karyotype according to ISCN (1978) (1); the technique used, if deviating from normal trypsin-Giemsa, and the number of cells analysed is also useful information. For example: 'normal male karyotype, 46,XY' would be a perfectly adequate report and it would imply an acceptable quality of the preparation. Any abnormal karyotype should be explained in plain writing, for example, '47,XX,+21, i.e., regular female trisomy 21 — Down's syndrome'. Complex situations should be sorted out by writing an explanatory letter containing the necessary advice for counselling and further action. A counselling appointment with the local clinical geneticist should be offered.

14. LABORATORY DATA MANAGEMENT

Data is a basic resource that is needed for the efficient running and control of day to day activities in a cytogenetics laboratory. Clearly, the value of data depends on the particular needs of the various users. In a big department, the names of outstanding specimens may be the only data of common value. Data that is frequently used must be easily accessible. Information sources should be structured accordingly. The human mind is a rotten store. It always plays on the theme: 'He who runs the information runs the show'. Ledgers, files, card indices and computers may sound dreadfully boring to the cytogeneticist, but they are obviously necessary for the cooperative handling of data on a larger scale. Even if the actual designs for the storage of data and the methods of information retrieval may vary immensely between different laboratories, the information that is needed in relation to a cytogenetics service must be very similar.

The minimum essential information has a very simple structure. All we need to know for each specimen that is received for chromosome investigation is the following.

(i)   *The identity of the patient.* That includes name, sex, date of birth, hospital number and address. Possibly a telephone number for contact in prenatal diagnosis cases.

(ii) *Who sent the specimen and to whom the results should be sent.* That includes hospital/GP and consultant.

(iii) *When it was taken and what type of tissue it was.* For example: 2 ml blood in lithium heparin taken 7.9.85 at 3.15 p.m.

(iv) *The reason for referral.* For example: mental deficiency.

These four pieces of information are usually duly filled in on the chromosome request card that accompanies the specimen. All we need to do as it arrives is to create a fifth piece of information, namely to write down the date received, to assign a case number, to decide whether it is an urgent specimen or not and to allocate it for processing and analysis. The referral card can then be stored in a systematic way whilst the specimen is being processed and analysed. The back page could be used for notations but it is usually better to keep a separate book. The sixth and final piece of information is the result which can be typed on to the request card as soon as they are available. The card can then be sent back as a report. A copy must be kept for filing purposes.

Although such a simple system is easily manageable the amount of information that can be retrieved is entirely dependent on how the results are stored. Sequential filing would be useless for any other purpose than annual statistics unless separate index registers were kept.

To provide the necessary documentation of day-to-day activities (e.g., for the purpose of being able to answer queries over the telephone), cytogenetics laboratories cannot avoid organising an information system of some sort. The actual design of these systems is probably not the same in any two laboratories. There is no best model, but some designs are better than others in a particular environment. However, the different elements of such information systems are probably rather similar and fairly well represented by the system that has developed over a number of years in our department. Let us briefly discuss its elements. The key for access to the results (i.e., the copy of the referral card), amongst a hundred thousand other letters, pedigrees, photographic negatives and so on, in our department is the case number. These are assigned as soon as specimens come in and can be found from an alphabetical register of index cards, or alternatively from the laboratory book. There is a separate ledger for blood samples, one for amniotic fluids and so on. All incoming specimens are entered in succession. Each culture also receives a separate laboratory number so that it will always be possible to tell from which flask a particular chromosome spread came. Information as regards who has been allocated the sample, media changes, harvest, analysis, and so on, is also entered into these books in a communal way. Therefore they are always kept up to date. These ledgers represent the most reliable data we have regarding the fate of incoming specimens. Similarly, the results are most accurately represented by the analysis sheets. The microscope slides and photographs represent the most important documentation and must be kept for a reasonable period. We store the abnormal slides indefinitely and the normal ones for at least 6 months. However, neither of these sets of data would contain sufficient information as regards patient's or doctor's addresses, because such data are usually only of value to the clinical and secretarial staff in the department. To obtain this piece of information we would have to find the case notes. These files contain much other information that is sometimes very relevant to the cytogenetic investigation. For example, the case number, or key to the file, is not unique to the individual but may refer to a whole line of ancestors and relations, sometimes found in cross-indexing of the referral cards as they come in for the assignment of a case number. It certainly gives another dimension to the cytogenetic investigation if it concerns a whole family with a chromosomal anomaly in some of its members, rather than being a struggle with these cases under the illusion that they represent isolated events. For the same purposes a separate card index is kept for all chromosomal abnormalities.

By necessity, this system also includes: lists of the addresses and telephone numbers of various doctors in the region, which must be kept up to date; weekly lists of amniocentesis clinics; follow-up of prenatal diagnosis cases, including lists of AFP results and abnormal scans (normal chromosome results must not be sent to patients who have had a termination of pregnancy for NTD); a copy of the referral card for back-up if the case notes go astray, which happens occasionally; a computer file for information retrieval, etc.

A key feature of this system is its distributed storage of intelligence. Part of the information is redundant in the sense that it is hardly ever used. Some of the more essential information is duplicated, much like the genetic material in a diploid organism;

perhaps it is a sign of evolution. Missing case notes can cause terrible havoc!

The information system that I have just described is more complex than the minimum essential model. Its principles are much the same as that of a computerised data base that provides vital information necessary for the quality and maintenance of a service, and which would have been missed if the conveyor-belt principle had been applied to the samples with no opportunity for information retrieval except annual statistics. Why are we not yet on a computer? The main reason is probably economical. It would be simple enough to fully computerise the cytogenetics laboratory service. It is more difficult to envisage a fully computerised communication between clinicians. They would have to keep their clinical letters in filing cabinets for the foreseeable future. Which computer could one opt for? There are computers of different sizes and designs. I do not think a home computer is a good option for a large cytogenetics laboratory. Data base facilities are essential. Furthermore, it would require a multiuser system that can be used interactively (e.g., when answering queries over the telephone), which means that it would have to incorporate all data that is presently continuously fed into laboratory books, analysis sheets and so on. Otherwise the computer may cause more trouble than it is worth.

15. CHROMOSOME REPOSITORIES

The Human Genetic Mutant Cell Repository at the Institute for Medical Research in Camden, New Jersey is probably the most widely known repository of cell cultures from patients with genetic diseases and chromosomal anomalies, although almost every major research laboratory in the field has a liquid nitrogen cell bank of their own for the storage of fibroblast cultures and lymphoblastoid cell lines from clinical cases with chromosome rearrangements. These cell banks usually also include panels of cell hybrids representing various normal and abnormal human chromosomes on a mouse or hamster background. The purpose of some of these collections is to provide chromosomes residing in a form that can easily be further propagated for use in molecular genetics for gene mapping and for the isolation of specific genes. The ultimate aim, of course, is the prevention of genetic disease.

There may be other materials that one also would like to include in a repository for the same purpose. For example, chromosome libraries of uncloned plasmid, phage or cosmid recombinant DNA of particular chromosomes, or parts thereof, are also suitable. Human X chromosome libraries have been available since the early 1980s. However, it would be far beyond the scope of this book to try to provide an account of all the possible uses of these materials. The present situation as far as clinical cytogenetics is concerned is that even if most workers in this field realise the importance of saving material from clinical cases, they only rarely have an opportunity to be involved in the future use of the abnormal chromosomes they have detected and characterised, which might detract from the continued and sustained interest in these activities. However, the situation may well change, for example, when *in situ* molecular DNA techniques become available that will allow simple direct localisation under the microscope of rearranged or deleted genes.

16. QUALITY ASSESSMENT

Clinical cytogenetics is a discipline that, contrary to general belief, probably involves as much theory as any other applied science and also has many of the attributes of craftmanship. The maintenance of high standards is essential, especially within the area of prenatal diagnosis. In 1982, the Association of Clinical Cytogeneticists (ACC) in the UK set up a working party to examine the possibility of establishing standards of analysis for diagnostic cytogenetic laboratories (15). The working party produced a scheme for the regular review and assessment of performance which concentrated on the following elements of routine diagnostic work.

(i) *Success rate*: from what proportion of specimens did a laboratory obtain a result?
(ii) *Reporting time*: how long did a laboratory take to provide a report to the clinician?
(iii) *Procedures/methods*: what procedures were used in making chromosome preparations?
(iv) *Preparation quality*: what was the quality of the preparation from which analysis was done?
(v) *Methods of analysis*: how many cells/colonies/cultures were scored and analysed? What methods were used for the analysis and checking of results?
(vi) *Reporting practice*: what form of reporting was used?
(vii) *Follow-up*: what arrangements are made to follow up tested pregnancies, to arrange extra tests or to follow up family members who are at risk?
(viii) *Mistakes*: how many mistakes are made?

Participating laboratories were invited to provide 1982 data from randomly selected days of the year anonymously. For example, let me condense the prenatal diagnosis data: the average amniotic fluid sample had a 94% success rate and was reported 22 days after the receipt of the specimen; it was a trypsin-Giemsa banded preparation and between 1 and 5 cells had been analysed and up to 12 cells counted; its chance of being followed-up was 72%; significant mistakes were very rare; some laboratories performed better than others, especially in producing preparations of high quality and quick reporting, but it is difficult to make certain that this also reflects a higher quality service over all.

The results, which perhaps were disappointing, seemed to improve slightly the following year, but the question of defining adequate standards largely remains an unsolved problem. Presumably, there are also other dimensions of diagnostic cytogenetic work of importance for the overall performance of individual laboratories and influencing those eight variables that were assessed (e.g., workload, obstetricians performance, professional judgement, funding, equipment, etc.). However, in spite of these difficulties, it seems fairly obvious that quality assessment schemes of this kind are feasible. They may be a controversial issue for other reasons.

17. ACKNOWLEDGEMENTS

I would like to thank all colleagues in the Department of Medical Genetics for much constructive criticism, Anne Naylor and Alison Towner for typing the manuscript and Linda Cheetham for the illustrations.

18. REFERENCES

1. ISCN (1978) *An International System for Human Cytogenetic Nomenclature (1978)*, Birth Defects: Original Article Series, Vol. **IXV**, No. 8, The National Foundation, New York.
2. ISCN (1981) *An International System for Human Cytogenetic Nomenclature — High-resolution Banding (1981)*, Birth Defects: Original Article Series, Vol. **XVII**, No. 5, March of Dimes Birth Defects Foundation, New York.
3. Schwartz.S. and Palmer,C.G. (1984) *Am. J. Med. Genet.*, **19**, 291.
4. Savage,J.R.K. (1977) *Nature*, **270**, 513.
5. Stene,J., Stene,E. and Mikkelsen,M. (1984) *Prenatal Diagnosis*, **4**, 81.
6. Baumgarten,A., Schoenfeld,M., Mahoney,M.J., Greenstein,R.M. and Saal,H.M. (1985) *Lancet*, **i**, 1280.
7. Ward,B.E., Henry,P.G. and Robinson,A. (1980) *Am. J. Hum. Gent.*, **32**, 549.
8. Schinzel,A. (1983) *Catalogue of Unbalanced Chromosome Aberrations in Man*, published by de Gruyter, Berlin and New York.
9. Lindenbaum,R.H., Hulte'n,M., McDermott,A. and Seabright,M. (1985) *J. Med. Genet.*, **22**, 24.
10. Ferguson-Smith,M.A. and Yates,J.R.W. (1984) *Prenatal Diagnosis*, **4**, 5.
11. Boue',A. and Gallano,P. (1984) *Prenatal Diagnosis*, **4**, 45.
12. Warburton,D. (1984) *Prenatal Diagnosis*, **4**, 69.
13. Hsu,L.Y.F. and Perlis,T.E. (1984) *Prenatal Diagnosis*, **4**, 97.
14. Hook,E.B. (1977) *Am. J. Hum. Genet.*, **29**, 94.
15. A.C.C. Quality assessment scheme in clinical cytogenetics, Survey of 1982 data, Report of the panel of advisers (unpublished).
16. Hook,E.B. and Hamerton,J.L. (1977) in *Population Cytogenetics: Studies in Humans*, Hook,E.B. and Porter,I.H. (eds.), Academic Press, New York, p. 63.
17. Boue',J., Nicholas,H., Barichard,F. and Boue',A. (1979) *Ann. Genet.*, **22**, 3.
18. de Grouchy,J. and Turleau,C. (1982) *Atlas des Maladies Chromosomiques*, Second edition, Expansion scientifique francaise, Paris.

CHAPTER 5

Diagnosis of Malignancy from Chromosome Preparations

C.J.HARRISON

1. INTRODUCTION

The idea that chromosome aberrations were the cause of the change from normal to malignant growth was first suggested by Boveri in 1914 (1). In 1960 the first consistent chromosome abnormality in human cancer was reported. This was the discovery of the Philadelphia (Ph[1]) chromosome in chronic myeloid leukaemia (CML) (2). Later, with the advent of chromosome banding techniques, this was found to be due to a translocation between chromosomes 9 and 22 [t(9;22)(q34;q11)]. Since that time an increased interest in the field of cancer cytogenetics has led to the discovery of a large number of specific chromosome abnormalities in association with certain neoplasms. These consistent chromosome changes are becoming increasingly important to clinical cytogenetics to aid in diagnosis, to provide an indication of patient prognosis and to monitor response to therapy.

In addition, a relationship has been demonstrated between the sites of some of these specific chromosome translocations and the position of cellular oncogenes. For example, it is now well known that the translocation between chromosomes 8 and 14 [t(8;14) (q24;q32)] associated with the majority of Burkitt's lymphoma, involves the c-*myc* oncogene. C-*myc* is located at the end of chromosome 8 and in affected cells c-*myc* is translocated onto the end of chromosome 14 (3). Other translocations involving 8q24 are also, less commonly, associated with Burkitt's lymphoma. These are t(2;8)(p11;q24) and t(8;22)(q24;q11). Chromosomes 2, 14 and 22, receiving the c-*myc* oncogene, also carry the immunoglobulin genes close to the translocation sites (4,5). This suggests that these loci may have a role to play in oncogenesis. The translocation t(9;22)(q34;q11) involves the exchange of the oncogenes c-*abl* and c-*sis* from chromosomes 9 and 22, respectively, to chromosomes 22 and 9 (6). This finding provided the definitive answer that the t(9;22) involved a reciprocal translocation. Previously it was uncertain whether any chromosome material was transferred from 9 to 22.

Therefore, in this area, cancer cytogenetics and molecular biology are closely associated and perhaps reflect the cancer genetic approach of the future. In spite of these important findings, the overall progress on cancer chromosome research has been slow and incomplete, owing to limited availability of tissue, a low yield of mitoses and the fact that mitotic chromosomes obtained from malignant cells are often contracted, 'fuzzy' and poorly banded.

Many of the basic cytogenetic techniques, as described in other chapters in this book, are used in cancer cytogenetics. However, most procedures have been modified in a variety of ways by individual laboratories to improve the mitotic yield and morphology of these difficult chromosomes. It would be impossible to describe all the modifications within the limits of this chapter. Therefore, I have chosen to describe here the most popular techniques, with those specific modifications most commonly used in this laboratory, that I have found to be most applicable in the provision of cytogenetic information for clinical use.

2. CULTURE TECHNIQUES

Those cases most commonly referred for cytogenetic investigation are the leukaemias. Samples of bone marrow and/or peripheral blood are studied for detection of chromosome abnormalities in the leukaemic cells. For the study of lymphomas, lymph node biopsy material is provided. It is rare to find abnormal lymphoma cells in the bone marrow of these patients. In patients with solid tumours a biopsy of the tumour itself is usually investigated, or serous effusions are examined directly. These commonly result from secondary malignancies or less commonly from lymphomas, pleural mesotheliomas or chronic myeloid leukaemia. The culture techniques used vary according to the type of tissue to be processed.

2.1 Culture Media

The medium used for samples of bone marrow, peripheral blood, lymph node or serous effusions is RPMI 1640 supplemented with 20% fetal calf serum, antibiotics; benzylpenicillin (sodium) B.P., 100 000 units per litre; streptomycin sulphate, 100 mg/l; glutamine 292.3 mg/l. This will be referred to as medium[1] in the text. It has been found to be the best for all cultures involving human lymphocytes or their precursors and the majority of suspension cultures. This medium, however, is not exclusive; other media or reduced serum levels are also successful.

The medium used for solid tumour material is Eagle's minimal essential medium (MEM) supplemented as above. This is a broad spectrum medium suitable for the growth of monolayer cultures.

L-15 medium supplemented as medium[1] may also be used as this has the advantage of growth selection for epithelial cells by retarding growth of fibroblasts. Where MEM or L-15 are used will be referred to as medium[2] in the text.

2.2 Handling of Material

All material from human malignancies should be handled as potentially infectious, using Class II laminar flow cabinets, sealed centrifuge buckets and so on. Readers should refer to the Howie Code of Practice (7) or Health and Safety Regulations for their own laboratory.

2.3 Culture of Bone Marrow

2.3.1 Direct and Short-term Culture

(i)   A bone marrow sample is collected by the haematologist and immediately transferred to a sterile tube containing $10-20$ ml of medium[1], with 10 units/ml of preservative-free heparin, at room temperature, for transport to the laboratory.

Samples may be transported in heparin alone. However, the presence of complete medium, maintained at no less than room temperature, and rapid transport of cells helps to preserve the mitotic activity.

(ii)   As soon as possible after arrival in the laboratory, count the cells in suspension in the medium. The most convenient method of counting the cells is to use a haemocytometer, carefully agitating the cells to disperse them evenly throughout the medium.

(iii)   Take 0.1 ml of suspension and dilute to 2 ml using a graduated pipette.

(iv)   With the coverslip firmly adhered to the haemocytometer slide, add one drop of the cell suspension to each side of the slide and count cells in the four outer corner squares.

(v)   Divide the total cell number by four to obtain the average number of cells in each square. This number is the number of cells $\times$ 10^4. Multiply this by 20 to correct for the initial dilution. This figure gives the number of cells per ml of the original suspension.

(vi)   If delayed in transit or transported without medium, wash the cells in unsupplemented medium[1] and transfer to fresh complete medium[1] prior to counting. This may help to stimulate cell division. To 'wash' the cells, transfer to a suitable centrifuge tube and top up the tube with unsupplemented medium[1]. Spin at 200 g for 5 min.

(vii)   Discard the supernatant, add fresh medium to the tube and agitate to resuspend the cells.

(viii) Adjust the final concentration of cells to 10^6/ml in 10 or 20 ml cultures, dependent on the cell number. Set up the cultures in small tissue culture flasks and incubate in the vertical position, at 37°C, gassed with 5% carbon dioxide. The use of flasks provides an increased surface area for gas exchange to maintain the correct pH. Alternatively, sterile 25 ml Universal tubes or 15 ml plastic centrifuge tubes with screw caps may be used with the tube placed at an angle of 60° (8).

(ix)   Since bone marrow exhibits an inherent mitotic activity it can be used directly for chromosome analysis. This makes possible an immediate karyotypic study of the cells dividing at the time of marrow aspiration. For direct cultures add 0.05 μg/ml of colcemid to the culture at the time of setting up, for 1 h, and then harvest the cells as described in Section 4.1.

(x)   The quality and quantity of mitotic cells in direct preparations are extremely variable and often difficult to analyse. This may be improved in many cases by short-term culture of cells. It is important to set up and investigate both direct and short-term cultures since in some samples overgrowth of the abnormal leukaemic cells by a normal clone may occur during short-term culture. Alternatively, other cases may depend on short-term culture to promote cell division of the leukaemic clones and hence disclose the chromosomal abnormalities. This has been demonstrated in acute promyelocytic leukaemia (M_3). Short-term culture reveals the characteristic translocation t(15;17)(q22;q11) present in the leukaemic promyelocytes (see Section 7.2). Direct cultures rarely show the translocation. For short-term cultures add 0.05 μg/ml of colcemid after 24 and 48 h of incubation, for 1 h. Short-term culture, in conjunction with this shorter incubation in colcemid, produces an increased mitotic yield with less contracted chromosomes. Harvest the cells as described in Section 4.1.

2.3.2 Methotrexate Synchronisation Procedures

Yunis (8) has demonstrated that the use of methotrexate (MTX) cell synchronisation makes it possible to achieve high quality preparations with an increased mitotic index and elongated finely banded chromosomes. In our experience the use of MTX has no detrimental effect on bone marrow cultures from CML or acute non-lymphocytic leukaemias (ANLL), but in certain cases of acute lymphoblastic leukaemia (ALL) the mitotic index is significantly decreased. MTX is used in treatment of some cases of ALL and may interfere with growth of the leukaemic cells in culture. We have encountered a few cases of ALL in which the MTX culture showed only normal cells on analysis, when the untreated cultures revealed an abnormal clone. Therefore, to prevent misinterpretation of results always examine an untreated culture in parallel with an MTX-treated one. We have also found that MTX does not significantly increase the length of the leukaemic chromosomes but improves metaphase spreading and chromosome morphology. This in turn improves the quality of the chromosome banding.

The MTX synchronisation technique for leukaemic bone marrows is a modification of the method applied to blood cultures as described in Chapter 2. It was first applied to bone marrow by Hagermeijer *et al.* (9) and further modified by Yunis (8). Synchronisation of bone marrow cultures is not as predictable as stimulated lymphocytes due to the variation in cell cycle length of the cells involved in different haematological disorders. Therefore, when establishing these techniques in different laboratories, experiment with a range of incubation times at each stage of the procedure until the optimum conditions are achieved.

(i)   Set up bone marrow cultures as described above (Section 2.3.1) and incubate at 37°C for $3-5$ h. The exposure of leukaemic cells to MTX without prior adaptation to the culture environment often results in unsuccessful growth.
(ii) Add MTX at a final concentration of 10^{-7} M and incubate at 36°C for 17 h.
(iii) Release the cells from the block by two washes with unsupplemented medium[1] and incubate again with 10^{-5} M thymidine for 6 h.
(iv) During the last $10-30$ min of the release, expose the cells to 0.05 μg/ml of colcemid at 37°C and harvest as described for bone marrow cultures (Section 4.1).

2.3.3 Fluorodeoxyuridine Synchronisation Procedure

The use of fluorodeoxyuridine (FUdR) for synchronisation of bone marrow cells for cytogenetic analysis was first described by Weber and Garson (10). In our experience the results achieved are similar to MTX synchronisation. FUdR is less toxic to the leukaemic cells and produces a higher mitotic index in cases of ALL.

(i)   Set up bone marrow cells as described above (Section 2.3.1) and incubate at 37°C for $6-8$ h.
(ii) Expose cells to FUdR at a final concentration of 0.1 μM together with 4 μM of uridine for 17 h. Excess uridine ensures that fluorouridylate, which can be formed from FUdR, is not incorporated into RNA (11).

(iii)   Add thymidine (10 μM) for 6 h to release the DNA synthesis block. No washing of cells is necessary, therefore cell loss is reduced and as a result a higher mitotic yield is obtained.

(iv)   During the last $10-30$ min of release add 0.05 μg/ml colcemid at 37°C. Harvest the cells as described for bone marrow (Section 4.1).

2.4 Culture of Leukaemic Blood

The examination of peripheral blood with and without phytohaemagglutinin (PHA) yields results of a different meaning. When PHA is added to the culture medium it produces division of the T lymphocytes. In cancer cytogenetics a PHA-stimulated culture is established in each case where there is the need to rule out the presence of a constitutional rather than a malignant cell chromosome defect. Unless leukaemic cells are present in the blood, usually no mitotic figures are found in the absence of PHA. When leukaemic cells are present in the peripheral blood the processing of this blood without PHA may reveal the basic karyotype of the condition, particularly in CML or acute leukaemia. This eliminates the taking of bone marrow for cytogenetic investigation only.

2.4.1 *Culture of Blood for Constitutional Karyotype*

The details of handling this type of culture are described in Chapter 2. In our laboratory the same medium[1] and culture techniques are used as for bone marrow samples.

(i)   Set up 1 ml of heparinised peripheral blood in 20 ml of medium[1] containing 0.2 ml of PHA in a small tissue culture flask incubated at 37°C in the vertical position.

(ii)   Culture for 72 h, then add 0.05 μg/ml of colcemid for 1 h prior to harvesting (Section 4.2).

(iii)   Alternatively, cells may be synchronised for 'high resolution' analysis. The technique presented here (8) represents an improved version of the original Yunis technique (12,13) since it incorporates the use of bromodeoxyuridine (BrdU) instead of thymidine to release the cultured lymphocytes from the MTX block. After 72 h of culture add MTX at a final concentration of 10^{-7} M to the flask.

(iv)   Incubate for a further 17 h then release the cells from the MTX block by washing twice with unsupplemented medium[1] and incubate at 37°C with $10-12$ μg/ml BrdU for 5 h.

(v)   During the last 10 min of this release expose the cells to 0.05 μg/ml of colcemid at 37°C prior to harvesting (Section 4.2).

2.4.2 *Culture of Unstimulated Blood with a Normal White Cell Count*

The setting up of unstimulated peripheral blood, (i.e., with no added mitogens) is dependent on the number of white cells. Frequently, when the white cell count is normal there is little, if any, mitotic activity in the unstimulated blood. However, this procedure is usually carried out if no bone marrow is supplied. If bone marrow has been provided it is of interest to culture the blood separately, as in a few cases a different karyotypic picture has been observed in the blood and marrow.

(i)   Set up whole blood as described for constitutional karyotype (Section 2.4.1), without the addition of PHA, and incubate at 37°C.

(ii)   Add 0.05 μg/ml of colcemid for 1 h for a direct culture, or add colcemid after 24 or 48 h incubation as described for short-term culture of bone marrow (Section 2.3.1).

(iii)   Harvest these unstimulated blood cultures as described for bone marrow (Section 4.1).

2.4.3 *Culture of Unstimulated Blood with a High White Cell Count*

If the white cell count is high it is usually an indication that leukaemic cells are present in the blood. It is advantageous to separate the white cells from the other blood cells.

(i)   Place 10 ml of lymphocyte separation medium into a sterile 25 ml Universal tube.

(ii)   Carefully layer 10 ml of fresh whole blood onto the surface of the separation medium by holding the tube at an angle and slowly running the blood down the side of the tube from a pipette.

(iii)   Centrifuge at 700 g for 20 min. After this time the red blood cells will have settled in a pellet at the bottom of the tube, leaving the white cells in a layer between the separation medium and the serum.

(iv)   Handle the tube gently to prevent disturbing the white cell layer and carefully remove the white cells with a siliconised sterile Pasteur pipette or a syringe with a wide bore needle. Take care to remove as little of the separation medium with the cells as possible.

(v)   Place the cells into a sterile centrifuge tube, make the suspension up to 10 ml with sterile phosphate-buffered saline (PBS) and centrifuge at 450 g for 10 min.

(vi)   Remove the supernatant, resuspend in fresh PBS and spin at 150 g for a further 10 min.

(vii)   Resuspend the cells in medium[1] and count.

(viii)   Set up 10^6 cells/ml in medium[1] in small tissue culture flasks.

(ix)   For bloods from patients with CML or acute leukaemia the raised white cell count is usually due to the presence of dividing blast cells. Therefore treat these cases as bone marrow samples following the procedure for direct and short-term bone marrow cultures (Section 2.3.1).

(x)   In the majority of cases of chronic lymphocytic leukaemia (CLL) the malignancy is of B cell origin, and the raised white cell count is due to an increased number of leukaemic B lymphocytes in the blood. These have a very low spontaneous mitotic index and are poorly stimulated by most mitogenic substances used previously, which are primarily T cell activators (e.g., PHA and pokeweed mitogen). A group of mitogens has been identified that primarily stimulate B cells. These polyclonal B cell activators will specifically stimulate CLL leukaemic cells (14) and include the following.

    (a)   Lipopolysaccharide from *Escherichia coli* 055, B5 (LPS). Add LPS to the cultures at a final concentration of 40 μg/ml.

    (b)   12-O-Tetradecanoylphorbol-13-acetate (TPA). Add TPA to the cultures at a final concentration of 10 ng/ml. Dissolve the stock solution of TPA in absolute ethanol and store at $-20°C$, dilute immediately before use in medium[1].

(c)   Supernatant from an Epstein Barr virus (EBV) producing permanent cell line B95-8. EBV has been found to be the most reliable and easiest method to use in our laboratory.

Keep a culture of B95-8 cells growing in the same medium[1] at a high cell density. Allow the culture medium[1] to become exhausted and of low pH by leaving the medium unchanged for $5-7$ days. This stimulates release of the virus and/or growth factor into the medium. Collect this supernatant from the culture. Spin at 300 g for 10 min to remove all cells. Filter this supernatant with a 0.2 μm filter. Divide the filtered supernatant into 2-ml aliquots and store at $-20°$C. Add a 2-ml aliquot to each 20-ml culture of separated CLL lymphocytes (1:9 of culture).

(xi)   Culture CLL cells with mitogen (a), (b) or (c) above; incubate at 37°C in a small tissue culture flask in the vertical position for 5 days.

(xii)   Harvest cells as described for PHA-stimulated blood (Section 4.2).

2.4.4 *Culture of Blood for Diagnosis of Fanconi's Anaemia*

There are three autosomal recessive gene disorders with an inherited predispositon to leukaemia and other forms of cancer; Fanconi's anaemia, Bloom's syndrome and ataxia telangiectasia. These syndromes are defined in more detail in Chapter 4. The increased risk of the development of malignancy in these patients has indicated the need for definitive diagnostic techniques.

Specific procedures can now be applied to blood samples from these individuals. In Fanconi's anaemia and Bloom's syndrome the results achieved are now regarded as highly diagnostic features. In ataxia telangiectasia the results are offered as an aid to the clinical diagnosis given by the neurologist or clinician involved.

The handling of Bloom's syndrome samples is described in Section 2.4.5 and ataxia telangiectasia in Section 2.4.6.

(i)   For diagnosis of Fanconi's anaemia, set up three cultures of peripheral blood stimulated with PHA for the suspected Fanconi's anaemia patient and a further three cultures from an age-matched control, as described in Section 2.4.1. It is preferable that the blood samples be collected at approximately the same time and treated in parallel throughout the procedure.

(ii)   Prepare a suitable stock solution of mitomycin C and add to two of the pairs of cultures, at the time of setting up, at final concentrations of 10 ng/ml and 50 ng/ml. Alternative cross-linking mutagenic chemicals may be used if preferred, for example, diepoxybutane. Reserve the third pair of cultures as untreated controls.

(iii)   Incubate the cultures at 37°C for 72 h, then add 0.05 μg/ml of colcemid for 1 h. Harvest as described in Section 4.2.

(iv)   Stain the slides with Giemsa as described in Section 5.1.

(v)   Score at least 100 cells from each culture type for the incidence of chromosome aberrations, in particular exchanges, dicentrics, rings and fragments. These configurations are defined in Chapter 4. The Fanconi's anaemia patient should show at least a 10-fold higher frequency of spontaneous chromosome aberrations in the untreated culture and of induced aberrations in the mitomycin C treated samples, than the normal control.

2.4.5 *Culture of Blood for the Diagnosis of Bloom's Syndrome*

(i)   Set up a PHA-stimulated peripheral blood culture from the suspected Bloom's syndrome patient and an age-matched control as described in Section 2.4.1.

(ii)   Add BrdU to these cultures and follow the procedure described in Chapter 3 (Section 4.7.3) for the detection of sister chromatid exchanges (SCEs).

(iii) Score the prepared slides for the incidence of spontaneous SCEs. Determine the mean SCE frequency from scoring 50 metaphases. The Bloom's syndrome individual will show a 10- to 15-fold increase in SCE frequency when compared with the normal control.

2.4.6 *Analysis of Lymphocyte Chromosomes as an aid to the Diagnosis of Ataxia Telangiectasia*

(i)   Set up five cultures of peripheral blood stimulated with PHA for the suspected ataxia telangiectasia patient and a further three cultures from an age-matched control, following the procedure described in Section 2.4.1.

(ii)   Incubate two of the cultures from each individual for 72 h and then add 0.05 μg/ml of colcemid for 1 h. Harvest as described in Section 4.2.

(iii) Stain the slides from one pair of cultures with Giemsa as described in Section 5.1.

(iv)   Score at least 100 cells from these slides for spontaneous incidence of translocations, dicentrics, exchanges, ring and inverted chromosomes. Ataxia telangiectasia individuals will more often shown an increase in these aberrations when compared with normal individuals.

(v)   Process the slides from the second culture of the suspected ataxia patient for G-banding, as described in Section 5.3. Karyotype as many cells as possible, as clones with rearrangements involving chromosomes 7 and 14 may be present.

(vi)   Incubate the three remaining pairs of cultures for 68 h at 37°C.

(vii) Expose one pair of cultures each with 50, 100 and 150 rads of X-rays, at 68 h and re-incubate for a further 4 h after irradiation. For the final hour of incubation add 0.05 μg of colcemid.

(viii) Harvest as described in Section 4.2 and stain with Giemsa as described in Section 5.1.

(ix)   Score 50−100 cells for the incidence of chromosome aberrations from each dose of X-rays. The 4 h incubation after irradiation ensures that those cells which were in G_2 at the time of irradiation have entered mitosis. The ataxia telangiectasia individual will show a 5- to 10-fold increase in chromatid-type damage when compared with the normal control.

(x)   It is important that the constitutional karyotype, and the frequency of both spontaneous and X-ray induced chromosome aberrations are considered together (Dr. A.M.R.Taylor, Department of Cancer Studies, Birmingham, personal communication).

2.5 Culture of Lymph Node Tissue from Lymphomas

It is now becoming evident that non-Hodgkin's lymphomas (NHL) can be divided into clinically significant subtypes depending on the chromosome defect involved. In most cases of Burkitt's lymphoma, for example, malignant cells have a translocation between chromosomes 8 and 14 [t(8;14)(q24;q32)]. Also, patients with follicular and diffuse histological subtypes of NHL can be identified cytogenetically (15). Cytogenetic study of lymphomas is best accomplished by combined analysis of a direct and a short-term culture of the affected lymph node. Direct preparations are preferable but lymphomatous tissue does not often contain sufficient numbers of dividing cells, and short-term culture stimulates proliferation. Consistent and successful culture of this tissue is more difficult to achieve than with leukaemic bone marrow, since death *in vitro* may be considerable.

In some lymphomas, particularly childhood disease, the patient presents with a leukaemic phase. In these cases blood or bone marrow can be examined. Use of bone marrow to study lymphomas in general has not been found suitable, since the small percentage of lymphoma cells found, in conjunction with their low mitotic index, often results in failure to find an abnormal karyotype.

2.5.1 *Direct and Short-term Culture of Lymph Nodes*

(i)   Bring the lymph node tissue from the operating room as soon as possible. It is unusual for a whole lymph node to be supplied; usually only a small biopsy is presented. It is advantageous if the surgeon will place the tissue directly into unsupplemented medium[1] to prevent drying out during transport.

(ii)   Transfer the tissue to a Petri dish containing $1-2$ ml of medium[1] and remove excess fat, blood, connective tissue or necrotic areas.

(iii)   Place the tissue inside a 'basket' of sterile gauze, in the Petri dish, and chop up the tissue with surgical knives. The function of the gauze is to remove all large clumps of cells and tissue debris, therefore the gauge is not critical. If the node is soft, cells will burst into the medium giving it a milky appearance.

(iv)   If the tissue is of harder consistency, slice or mince the material into small segments and release the cells by pressing the pieces of tissue against the gauze, using the plunger of a sterile syringe. This will release cells into the medium[1].

(v)   Collect the cells in suspension into a sterile centrifuge tube in unsupplemented medium[1] and top up to 10 ml.

(vi)   Centrifuge at 200 *g* for 8 min and resuspend the cells in fresh complete medium[1].

(vii)   Count the cells and determine their viability using an exclusion dye, for example, trypan blue in the ratio of 50% dye to 50% cell suspension. This step is important as many cells are damaged during the mincing process.

(viii) Set up 10^6 viable cells per ml in 20 ml of medium[1] in each T30 tissue culture flask, as for bone marrow direct and short-term culture (Section 2.3.1), and incubate at 37°C.

(ix)   For direct preparations add 0.05 µg/ml of colcemid at the time of culture initiation and incubate for 1 h. Add colcemid to short-term cultures after 24 or 48 h incubation. Harvest the cells as described for bone marrow (Section 4.1).

(x)   Set up MTX and FUdR synchronised cultures as described for bone marrow samples (Sections 2.3.2 and 2.3.3, respectively). Harvest, also as described for bone marrows (Section 4.1).

(xi)   Often direct and short-term cultures yield poor quality, poorly spread metaphases. Also, some B type differentiated lymphomas require B cell mitogens for cell proliferation. Therefore, since diagnosis is often unknown prior to cytogenetic investigations, it is advantageous to set up lymph node cultures with B cell mitogen added, as described for cultures of CLL lymphocytes (Section 2.4.3).

(xii) Culture for 5 days and harvest as described for PHA-stimulated peripheral blood (Section 4.2). This stimulated culture will produce metaphases of improved quality, but it is essential to consider the results in conjunction with direct or short-term cultures, since evolution or selection of clones may arise over the longer culture period.

2.6 Culture of Solid Tumour Material

Cytogenetic analysis of solid tumours often produces technical problems. Malignant tumours are often characterised by a low mitotic index and disaggregation techniques frequently yield a large number of non-viable or non-dividing cells. Lengthened colcemid treatment usually has to be applied to obtain sufficient numbers of metaphases, resulting in highly contracted chromosomes. The techniques described here provide a number of different conditions suitable for a variety of tumour types. For example, better growth is obtained from tumours of the prostate by explanting, whereas bladder tumour cells respond better to disaggregation by collagenase.

2.6.1 Cell Suspension Technique

(i) Bring solid tumour specimens from the operating theatre, in a sterile container, as soon as possible. If delays are envisaged, the surgeon should place the tissue into a sterile tube containing medium².

(ii) On arrival in the laboratory place the tissue in a sterile Petri dish containing $1-2$ ml of medium² and remove bloody, necrotic and normal tissues.

(iii) Cut the tumour tissue into small pieces with sterile surgical blades. Attempt to obtain precise smooth cuts to avoid tearing the tissue.

(iv) If the tumour is soft or semi-solid in consistency, agitate the minced tissue in the medium², which will release sufficient numbers of viable tumour cells for cytogenetic analysis.

(v) To remove the larger pieces of tissue push the supernatant through a sterile gauze (see Section 2.5.1) and wash the gauze a couple of times to collect all of the released cells.

(vi) Count the cells and determine viability by staining with trypan blue, as described for lymph node cells (Section 2.5.1).

(vii) If the cell number or viability is low, or if the tumour is of a solid nature, treat the small pieces of tissue with freshly made 0.8% collagenase I, II or IV and 0.002% deoxyribonuclease I in medium², previously sterilised by filtration. Collagenase is used as it has no effect on the cells but destroys the connective tissue, liberating the tumour cells from their fibrous stroma. Collagenase alone causes the tissue to become gelatinous. Deoxyribonuclease is added to prevent this adverse effect.

(viii) Place the tissue into the enzyme mixture and incubate for 30 min to 1 h at 37°C.

(ix) Transfer the cell suspension, without the remaining tissue fragments, into a sterile disposable centrifuge tube. Removal of tissue is best done by passing the suspension through a sterile gauze. Gently force through as much tissue as possible using the sterile plunger of a syringe. Rinse the tissue 2 or 3 times with additional medium[2] to collect all the released cells and to inactivate the enzyme.

(x)   Wash the cells twice by centrifugation at 200 *g* for 8 min followed by resuspension in fresh complete medium[2].

(xi)   Count the cells in suspension in the medium[2] as in step (vi).

(xii) Transfer $3-5 \times 10^6$ cells/ml from step (vi) or (xi) to a small tissue culture flask in 5 ml of medium[2] and incubate horizontally, with the cap loosened, at 37°C in a 5% CO_2 gas incubator.

(xiii) Alternatively, cells may be grown on a feeder layer of fibroblasts. Tumour cells often grow more efficiently on feeder layers. The most effective are X-irradiated fibroblasts, since they provide attachment and nutrients for tumour cells but are unable to divide themselves.

(xiv) After 1 or 2 days in culture examine the flasks with an inverted microscope to assess growth. Tumour cells are usually of epithelial origin, varying in size and shape dependent on the type of tumour. They are usually large and tend to grow in clumps either in suspension or loosely attached to the feeder layer or flask surface.

(xv)   Change medium[2] in flasks when a large number of tumour cell colonies are observed. This is usually between 5 and 9 days. If cells are in suspension, remove the supernatant, centrifuge at 200 *g* for 8 min, add fresh medium[2] to the cell pellet and plant the suspension back into the original flask, as cells may have been left in contact with this flask surface.

(xvi) Two days later add 0.05 μg/ml of colcemid for $2-4$ h. 0.015 μg/ml of colcemid may be added for up to 16 h in very slow dividing tumours. This, however, tends to produce highly contracted chromosomes.

(xvii) Alternatively, prior to addition of colcemid, the cells may be synchronised using the MTX or FUdR procedures, as described for bone marrow (Sections 2.3.2 and 2.3.3, respectively).

(xviii) Harvest the cells as described in Section 4.3.1.

2.6.2 *Tissue Explant Technique in Flasks*

(i)   Mince the tumour tissue as described for the cell suspension technique (Section 2.6.1) into small pieces $1-2$ mm³ in size.

(ii)   Place the pieces directly onto the bottom surface of a small tissue culture flask in a drop of medium. Set up a minimum of three flasks.

(iii) Incubate undisturbed overnight at 37°C in 5% CO_2 in a humid environment and allow to adhere to the surface of the flask without drying out.

(iv)   The following day add 5 ml of medium[2] carefully to prevent removal of tissue already adhered to the surface of the flask. Some pieces of tissue may still be floating in the medium but should not be removed as they may adhere after a further incubation.

(v)   Change the medium each week and observe growth. Fibroblasts will tend to grow out of the explants initially ($1-2$ weeks) which will serve as an autologous feeder layer for the tumour cells. After $2-3$ weeks rounded epithelium-like cells will migrate from the explant and begin to invade the fibroblast layer.

(vi)   Change the medium each week until the epithelial colonies become large enough for harvesting. When changing the medium take care not to disturb the tumour cells since they are often loosely attached.

(vii)   When the cells are ready for harvesting replace the medium 24 or 48 h prior to adding 0.05 μg/ml of colcemid to the culture for 4 h. Harvest as described in Section 4.3.1.

2.6.3 *Tissue Explant Technique on Coverslips (The Sandwich Technique)*

(i)   Mince the tumour tissue into small pieces, as described in Section 2.6.2.

(ii)   Place a 2 cm² square coverslip into the bottom of a small Petri dish.

(iii)   Place 3 − 4 pieces of tumour tissue, with straight edges, onto the coverslip, each in a tiny drop of medium.

(iv)   Place a second coverslip on top of the first at an angle, to make a sandwich. Press lightly and the two coverslips will seal together with the tissue and a small amount of medium between.

(v)   Add 2 ml of medium² to the Petri dish.

(vi)   Change the medium each week and observe growth. When considerable growth has occurred separate the two coverslips and transfer each to new Petri dishes with the cells uppermost. This provides two cultures from the same explant as cells will have grown on both the upper and lower glass surfaces.

(vii)   Add medium to both coverslips and to the original Petri dish, as cells may have overgrown from the coverslips onto the surface of the plastic if tissue culture grade Petri dishes are used.

(viii)   Allow further growth to continue until the cells are ready to harvest. Change the medium 24 or 48 h prior to harvesting and add 0.05 μg/ml of colcemid for 4 h.

(ix)   Harvest the cells on the coverslips as described for *in situ* harvesting (Section 4.3.2) and harvest the cells on the Petri dish as described in Section 4.3.1.

2.7 Culture of Established Cell Lines for Cytogenetic Investigation

The culture of tumour cells, in some cases, shows sufficient growth to give rise to established cell lines. More often, however, cell lines are established with the application of special culture techniques. It is inappropriate to go into detail of the establishment of such cell lines in this chapter. I will discuss only the methods of handling them once they have become established.

One problem to bear in mind with established tumour cell lines is that they may or may not resemble the original karyotype, since tumour cells *in vitro* often show secondary chromosomal changes as early as 6 − 10 passages. In spite of this disadvantage, established tumour cells are often used in cancer research and chromosomal characterisation is used along various stages of cultivation for assessment of results.

A wide variety of conditions are required for growth of tumour cell lines; for example, some grow better in suspensions, others in monolayers, whereas some will grow only with a certain cell density range and specific medium. This topic is too specialised for consideration in this chapter but certain standard conditions apply for cytogenetic analysis.

(i)   To obtain a good mitotic index use logarithmically growing cells, with the medium changed 48 h prior to harvesting.

(ii)   At this point, expose the cells to 0.05 μg/ml of colcemid for 1 h and harvest as described for short-term cultures of solid tumours (Section 4.3.1). With this simple approach it is possible to obtain an adequate number of early and mid-metaphases of good morphology. The resolution achieved is of sufficient quality to delineate the origin of marker chromosomes often found in tumours and tumour cell lines, which may not have been possible to decipher from the original tumour.

2.8 Culture of Serous Effusions

Cytologists often encounter difficulties in diagnosing malignancy in serous effusions. It is the less common causes of effusions which present the greatest problems, including lymphomas, pleural mesotheliomas or CML (16). In known leukaemias or lymphomas with effusions it is of interest for the clinician to know whether malignant cells are present in the aspirate, since chromosome aberrations are often observed in malignant effusions. One is usually analysing a secondary malignancy with late tumour progression when effusion is present. The greater the polyploidy the easier the cytologist finds the diagnosis. It is in those cases where the tumour is less varied, for example mesothelioma, that equivocal cytology results are found and it is in these cases that cytogenetics is of most value. The presence of a clone of cells with an unbalanced, abnormal karyotype indicates a malignant clone in such effusions. The direct culturing of effusions provides the best results for cytogenetics (16).

(i)   The patient should be moved around prior to aspiration to ensure that cells are in suspension in the fluid.

(ii)   Transfer the effusion directly from the patient into heparin, in a suitable sterile container, and transport rapidly to the laboratory.

(iii)   Incubate immediately on arrival at 37°C and aliquot approximately 10^6 cells/ml into 20 ml of fluid in a small tissue culture flask. Alternatively, the cells may be resuspended in a mixture of 50% fluid and 50% medium[2]. Complete removal of the fluid greatly decreases the mitotic index.

(iv)   The highest mitotic index is achieved by harvesting immediately. To do so, add 0.05 μg/ml of colcemid for 1 h at 37°C prior to harvesting.

(v)   Harvest the cells as described for direct or short-term cultures of bone marrow (Section 4.1).

(vi)   At the expense of a high mitotic index, better chromosome preparations may be achieved by 24 h culture prior to the addition of colcemid, or by the use of MTX or FUdR synchronisation, following the procedure as described for culture of bone marrow samples (Sections 2.3.2 and 2.3.3).

3. PRESERVATION OF BONE MARROW AND WHOLE BLOOD FOR CHROMOSOME ANALYSIS

During the course of chromosome analysis on patients with haematological disorders it is sometimes necessary to set up additional cultures. Samples stored in the refrigerator for long periods degenerate and fail to grow in culture. A need is growing for long-term preservation of bone marrow and blood samples in such cases as listed below:

(i)   Repeated samples from the same individual are difficult or impossible, often because the first sample is taken at diagnosis prior to treatment. Subsequent samples of the patient's blood or bone marrow would be depleted of leukaemic cells. Or

(ii) Repeated examination is often required to check a translocation or abnormality, or if the original culture failed for any reason. On a second culture high resolution techniques may be carried out which were perhaps not possible at the time of sampling.

A simple technique to carry out the preservation of blood or bone marrow cells has been developed (17,18).

(i)   Take 1 ml of heparinised whole blood or 2×10^6 lymphocytes, separated as described in Section 2.4.3, or up to 2×10^7 cells from a bone marrow sample, and suspend in 1 ml of medium[1] (or 1 ml of fetal calf serum) and 1 ml of 10% dimethyl sulphoxide (DMSO). Sterilise the DMSO prior to addition to the cells, by filtering through a 0.2 μm filter. DMSO has been found to be superior to glycerol for the freezing of blood cells.

(ii) Seal the suspension in a 2-ml screw cap plastic freezing vial and store in the vapour phase of a liquid nitrogen freezer for up to 1 year.

(iii) When required, thaw the cells rapidly in a water bath at 37°C.

(iv) Wash the cells in medium[1] by centrifugation at 200 g for 5 min and resuspend in fresh medium[1].

(v)   Culture the cells as previously described (bloods, Section 2.4, bone marrow, Section 2.3).

4. PREPARATION OF CULTURES FOR CHROMOSOME ANALYSIS

4.1 Preparation of Bone Marrow, Unstimulated Peripheral Blood, Unstimulated Lymph Node Cells and Pleural Effusions

The same harvesting technique is applied to direct and short-term cultures of bone marrow, unstimulated peripheral blood, unstimulated lymph node cells and pleural effusions. The procedure described here is the one we have found to be most universally successful when handling a variety of samples in a busy clinical laboratory.

(i)   At the end of the colcemid exposure time, transfer the cells in suspension in the medium into a centrifuge tube and spin at 200 g for 10 min.

(ii) Resuspend the pellet of cells in freshly made 0.075 M potassium chloride (KCl) for 15 min at room temperature. To continue the accumulation of metaphases during the hypotonic treatment, 0.05 μg/ml of colcemid may be added to the KCl. The optimal period of hypotonic is variable depending on the type of cell. In most cases it is recommended that the cells be exposed to hypotonic for no longer than 20 min, since a prolonged hypotonic treatment sometimes results in an ill-defined morphology. Problems are encountered with the exposure of leukaemic bone marrow samples as the chromosomes are frequently of 'fuzzy' and ill-defined morphology, but a longer incubation in hypotonic is required to facilitate the spreading, at metaphase, of these often tightly packed chromosomes.

The temperature of the hypotonic does not appear to be critical. Incubation of cells in KCl at either room temperature or 37°C produces the same result, in my experience. However, it is advisable not to heat the KCl to a temperature greater than 37°C or to use chilled hypotonic, as drastic changes in temperature may affect the physiological functions of the cell, resulting in inefficient action of the hypotonic effect.

(iii) Centrifuge the cells in KCl at 200 g for 5 min and resuspend the pellet of cells in 1 − 2 ml of freshly prepared 3:1 methanol:acetic acid fixative. This should be added over a few seconds, drop by drop, with constant agitation of the tube to ensure proper and fast fixation.

(iv) Adjust the volume to 10 ml and leave at room temperature for 20 − 30 min.

(v) Wash the cells in freshly prepared fixative an additional 2 − 3 times, resuspending the cells each time.

(vi) Store in the refrigerator at 4°C overnight.

(vii) Change the fixative an additional three times to eliminate cell debris and cytoplasmic background.

(viii) Many methods are routinely and successfully used to spread metaphases onto glass slides. To maximise the spreading of mitoses from these types of culture the slides must be pre-cleaned. Wet the clean slides in cold, distilled water and drop the chilled suspension from a height of about 4.5 feet onto slides placed at an angle of 30°, to overcome the resistance of the cell membranes. Alternatively, drop the chilled suspension onto the wet slides from a few inches away and follow this drop immediately by a drop of fresh cold fixative alone, from the same position.

(ix) Evaporate the fixative rapidly from the cells by blowing gently across the surface of the slide for a few seconds, taking care not to deposit moisture from the breath onto the slide, or dry the slide on a 37°C hot plate. Flaming of slides or the use of a 60°C hot plate are alternative methods used to maximise spreading of difficult cells. These rapid drying techniques tend to produce over flattened chromosomes which are difficult to G-band. Better results are achieved by using lower temperatures and greater height or extra fixative when dropping cells onto slides.

(x) To obtain optimal results with chromosome banding techniques desiccate the slides at room temperature for 3 − 14 days. Freshly made slides of leukaemic metaphases or the mitoses of malignant cells will not band well.

4.2 Preparation of Blood Cultured with PHA

This topic is covered in detail in Chapter 2. I will briefly describe here the technique used in this laboratory to coordinate the preparation of these blood cells with those obtained from malignant tissues.

(i) After colcemid exposure, transfer the cells in medium to a centrifuge tube and spin at 200 g for 8 min.

(ii) Resuspend the cells in freshly made 0.075 M KCl and incubate for 10 min at room temperature.

(iii)   Spin for 5 min at 200 *g* and resuspend in fresh 3:1 methanol:acetic acid fixative. Add the fixative drop by drop whilst agitating the cell suspension continuously.

(iv)   Leave the cells in this first fixative for 20 – 30 min at room temperature, then change the fixative six times more.

(v)   Make slide preparations by dropping the cells, resuspended in a dilute concentration of fixative (6:1, methanol:acetic acid), from a height of approximately 3 feet, onto wet slides placed at an angle of 30°. Alternatively, drop the cells in fresh 3:1 fixative onto the wet slides followed immediately by a drop of cold fixative alone, as described in Section 4.1.

4.3 Preparation of Tumour Cells

4.3.1 *Suspension Harvest for Tumour Cells*

Prepare tumour cells cultured by the cell suspension or explant techniques for flasks (Section 2.6.1 and 2.6.2, respectively) by the following method.

(i)   Following colcemid treatment, remove the cells from the surface of the flask or Petri dish by treatment with 0.05% trypsin and 0.02% EDTA solution for 5 min at 37°C.

(ii)   Remove the remaining cells from the bottom of the flask or dish with a rubber policeman.

(iii)   Resuspend the cells in fresh medium[2] containing serum and 0.05 µg/ml colcemid to inactivate the enzyme and to continue the accumulation of cells in metaphase at the same time.

(iv)   Centrifuge at 200 *g* for 8 min to remove the medium and resuspend in 0.075 M freshly made KCl, as hypotonic, for 15 min at room temperature.

(v)   Spin again at 200 *g* for 8 min and add freshly made 3:1 methanol:acetic acid fixative to the cells carefully without disturbing the pellet. Tumour cells, after trypsinisation, have fragile cell membranes. Fixation in a pellet protects the cell membrane from premature rupture and resultant loss of chromosomes from the metaphase cells.

(vi)   Allow the fixed pellet to stand at room temperature for 30 min to ensure complete fixation of the cells in the centre of the pellet.

(vii)   Centrifuge at 200 *g* for 8 min to collect those cells which may have been disturbed from the pellet, and add fresh fixative.

(viii) Change the fixative three times more and refrigerate overnight.

(ix)   For the slide-making procedure follow the steps in Section 4.1.

4.3.2 *In Situ Harvest for Tumour Cells*

Prepare cells cultured by the tissue explant technique for coverslips (the Sandwich Technique, Section 2.6.3) by the following procedure. The advantage of this technique is that the colonies grown up from individual pieces of tumour tissue are not disrupted by the harvesting procedure. Therefore, a cytogenetic comparison can be undertaken between pieces of tissue from different parts of the tumour.

The disadvantage of the technique is that the metaphases remain in close association with the non-dividing cells and this may reduce chromosome spreading, making chromosome banding analysis more difficult.

(i)   Remove the medium from Petri dishes containing coverslips with a Pasteur pipette and add 1 ml of fresh 0.075 M KCl for 20 min at room temperature.

(ii)   Add 1 ml of cold fresh 3:1 methanol:acetic acid fixative to each dish, into the KCl. This prevents shearing of the cells from the glass surface.

(iii)   Replace with 2 ml of fresh cold fixative and refrigerate at 4°C for a minimum of 30 min.

(iv)   Change the fixative again and remove the coverslips. Drain the excess moisture from the coverslips by holding at an angle of approximately 60° onto blotting paper.

(v)   Dry the coverslips on a 37°C hot plate and band them in the same way as slides made by the other harvesting methods.

5. BANDING TECHNIQUES

The details of chromosome banding techniques are described in Chapter 3. I will outline here those we have found the most successful in our laboratory for the banding of leukaemic and other malignant chromosomes.

5.1 Solid Staining

In general, chromosomes from leukaemias and other malignancies require a shorter staining time than longer term or stimulated cultures.

(i)   Stain the slides in a coplin jar in 2% Giemsa in Gurr's buffer (pH 6.8) for 3−4 min at room temperature. Stain longer term or stimulated cultures for 8 min.

(ii)   Rinse twice in tap water and carefully blot dry. Do not leave to air-dry as this swells the chromosomes and accentuates the ill-defined morphology characteristic of malignant cells.

(iii)   Examine the slides without mounting. Since malignant chromosomes vary greatly in their uptake of stain, this allows the opportunity to destain if the chromosomes are overstained or to add more stain if understained.

(iv)   If overstained, destain by passing the slides for 2 min each through $2 \times 70\%$ ethanol, $1 \times 90\%$ ethanol plus 1% hydrochloric acid, $1 \times$ absolute methanol and air-dry. This destaining procedure will also remove immersion oil from the slides without the need for xylene. Xylene has been found in some cases to deposit a film of opaque material onto the surface of the slides which is difficult to remove.

(v)   If understained, repeat steps (i) and (ii).

A second advantage of examining the slides unmounted is that many malignant cells have a very low mitotic index and only one or two good quality metaphases may be found on several slides. These metaphases, initially identified on solid stained slides, may be destained, as above, and subsequently G-banded.

5.2 Protein-coating

The chromosomes of malignant cells, especially the lymphoid malignancies, are often 'fuzzy' and do not take up stain very well after G-banding pre-treatments. To improve the staining ability and give increased definition to the G-bands, in this laboratory slides are protein-coated.

(i)   Dip the slides in freshly made 2% fetal calf serum in 3:1 methanol:acetic acid fixative.

(ii)   Drain the slides by resting on one edge and leave them to dry for a minimum of 2 h or overnight.

5.3 G-Banding

Two methods of G-banding have been found to be most successful for the G-banding of malignant cells in this laboratory and are described below.

5.3.1 *G-Banding Using Wright's Stain*

The best results are obtained with this technique on elongated pro-metaphase chromosomes, produced by the MTX (8) or the FUdR (10) synchronisation procedures described previously (Sections 2.3.2 and 2.3.3, respectively). Frequently, chromosomes of malignant cells are highly contracted, even when the above techniques are applied. Therefore, in these cases it may be necessary to use an enzyme treatment in addition to Wright's stain as described here. The advantage of this technique is that successive staining and destaining may improve the quality of the banding.

(i)   Make up a 0.25% stock solution of Wright's stain by dissolving 2.5 g of stain in 1 l of anhydrous methanol, and incubate at 37°C for 16 h.

(ii)   Age this solution for a minimum of 7 days at room temperature and filter prior to use.

(iii)   Mix one part of the stock solution with three parts of 0.06 M Sorensen's phosphate buffer (pH 6.8).

(iv)   Pour the diluted stain immediately onto the protein-coated slides for 2 – 3 min.

(v)   Rinse in tap water and blot dry.

(vi)   The chromosomes of the malignant cells may appear 'fuzzy' and poorly stained but may improve in sharpness by one or two successive destaining and restaining procedures. To destain, follow the method described in Section 5.1 (iv) above and then repeat steps (iii) to (v) in this section.

(vii)   For cases with highly contracted chromosomes dip the slides for 1 – 2 sec in 1% trypsin in Sorensen's buffer (pH 6.8) after destaining.

(viii) Rinse in buffer then stain following steps (iii) to (v).

5.3.2 *G-Banding Using Giemsa Stain*

This technique involves a modification of the acetic-saline-Giemsa (ASG) technique (19) and the trypsin method of Seabright (20). This method, in our laboratory, is applied to all types of malignancy to produce banding of consistently high quality.

(i)   Place the protein-coated slides in 2 × SSC in a water bath at 60°C for 1 h.

(ii)   Remove the slides and rinse in Hanks' HBSS solution pH 6.8.

(iii)   Place in 0.02% trypsin in HBSS (pH 8) for 5 – 10 sec.

(iv)   Rinse in HBSS pH 5.0.

(v)   Stain in 2% Giemsa for 3 – 5 min.

5.4. **Other Banding Techniques**

G-banding is the most successful and most frequently used banding technique for chromosomes of malignant cells. The other banding procedures may be carried out in some difficult cases, for example, to determine the origin of marker chromosomes or if normal polymorphisms are present. These alternative banding procedures include C-banding, Q-banding and R-banding, which are carried out in the routine manner as described in detail in Chapter 3. Therefore, they will not be further described in this chapter.

6. GUIDELINES FOR CHROMOSOME ANALYSIS

The majority of the labour in the handling of leukaemias and other malignant tissues is in the chromosome analysis. Even when the utmost attention has been applied to culture and preparative techniques, the mitotic index may be low and many slides may need to be scanned to obtain sufficient metaphases. Also, the chromosome morphology may be poor, with 'fuzzy' ill-defined chromosomes which refuse to respond to G-banding.

(i)   Mount permanent preparations of banded and non-banded slides with a suitable mountant.

(ii)   Analyse as many cells in detail as possible by light microscopy. A minimum of 20 metaphases is preferable. In cases with a low mitotic index analyse all available cells. In those cells with poor morphology, contracted chromosomes or reduced spreading full karyotypic analysis may not be possible, therefore count the chromosomes to provide an indication of numerical abnormalities or the presence of marker chromosomes.

(iii)   At the microscope inadvertent selection of superior metaphases by the analyst is inevitable. Therefore, try to select metaphases randomly, bearing in mind that the leukaemic or malignant cells will be of inferior morphology.

(iv)   Analyse a minimum of 30 metaphases if all normal cells are found, to rule out the presence of an abnormal clone. In some marrows with a preponderant percentage of leukaemic cells it is still possible to observe an essentially diploid picture due to the low mitotic index of the leukaemic or malignant cells compared with a high mitotic index of the normal elements. The incidence of abnormal cells observed will depend on the total number of metaphases examined. Obviously, the larger the number of cells examined the higher the chance of finding cells that deviate from the normal.

(v)   Photograph and karyotype as many cells as necessary, dependent on the quality of the metaphases and the complexity of the chromosome abnormalities to be identified. The dependence on karyotyping, especially with poor quality preparations and/or high resolution analysis, is far greater with cancer cell chromosomes than any other type of analysis. For example, in some solid tumours analysis by microscope may be impossible due to the inferior quality of the metaphases, therefore in these cases karyotype all metaphases. Also, karyotyping may be necessary to identify primary changes and consistent chromosome abnormalities from secondary or evolutionary changes.

(vi)   Prepare karyotypes on a selection of cells when the results appear to be normal, as minor chromosome abnormalities may be revealed when photographed chromosomes are cut out and laid side by side.

7. INTERPRETATION OF RESULTS

The cytogenetics of leukaemias and lymphomas has been closely correlated with diagnosis and prognosis. This chapter, in a book of practical clinical cytogenetics, is not the place to enter into too much detail but I will provide examples and refer the reader to repositories where this information is given in greater detail.

7.1 Chronic Myeloid Leukaemia

Specific translocations have been found to correlate with specific types of leukaemia. The best and most widely known example is that of the Philadelphia (Ph[1]) chromosome. This translocation between chromosomes 9 and 22 [t(9;22)(q34;q11)] is diagnostic for CML. Variants of the Ph[1] chromosome also exist, in association with this disease. Cases of CML in which no Ph[1] chromosome are observed are now thought to represent a different leukaemia type of Ph[1]-negative CML. The reader should refer to Sandberg (21) for discussion of the Ph[1] chromosome and to Mitelman (22) for details of all variant Ph[1] chromosomes and Ph[1]-negative cases.

Cytogenetics provides an early and reliable indication of the onset of blastic crisis in patients with CML. In the majority of these cases additional chromosomes are observed in the dividing cells of the bone marrow (*Figure 1*). The most common extra chromosomes are a second Ph[1], a chromosome 8 and/or an iso 17 chromosome. These findings are also described in Sandberg (21) and detailed in Mitelman (22). Therefore, cytogenetics may be used to monitor patient progress.

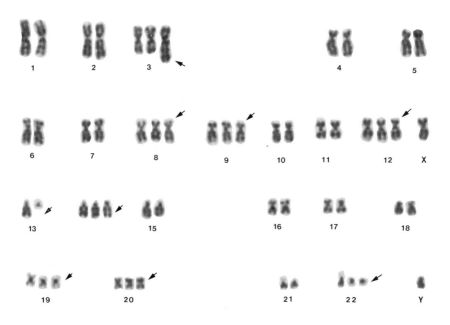

Figure 1. Karyotype from the bone marrow of a patient with CML in blastic crisis. An extra Ph[1] chromosome is present, with additional chromosomes 3, 9, 10, 12, 14, 19 and 20 and a translocation between chromosomes 3 and 13.

7.2 Acute Non-Lymphocytic Leukaemia

More recently the collection of cytogenetic data on patients with ANLL has revealed the association of specific chromosome translocations with subtypes of ANLL, in conformation with the French American British (FAB) morphological classification. These findings are described in detail in the Fourth International Workshop on Chromosomes in Leukaemia (23). Briefly, the FAB subtypes M_1 to M_6 show consistent chromosome changes. For example, 93% of all ANLL patients with t(8;21)(q22;q22) (*Figure 2*) belong to the M_2 group and the remaining 7% are M_4. t(15;17)(q22;q12) is diagnostic of M_3 (acute promyelocytic leukaemia) and has not been found outside this group. An inversion or deletion of chromosome 16 is diagnostic of eosinophilia in association with M_4, and deletions of chromosome 11q are frequently associated with M_5. Other chromosome abnormalities are generally found in association with all subtypes of ANLL, these include additional chromosomes 8 or 21 and the complete loss or deletion of the q arms of chromosomes 5 and 7. These findings and the more specific translocations also provide some indication of patient prognosis. For example, the presence of an extra chromosome 21 or the t(8;21) indicates a good prognosis, whereas abnormalities or loss of chromosomes 5 and 7 suggest a poor patient prognosis.

Monosomy of chromosomes 5 or 7 and abnormalities of chromosomes 3 and 17 are preferentially involved in secondary leukaemias (24). The majority of these are related

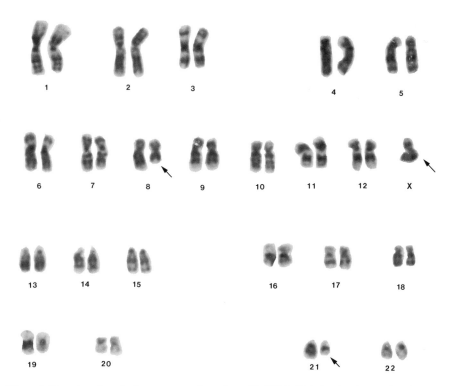

Figure 2. Karyotype from the bone marrow of a patient with ANLL (M_2) showing the translocation t(8;21) (q22;q22). This is frequently associated with the loss of a sex chromosome, as shown in this karyotype.

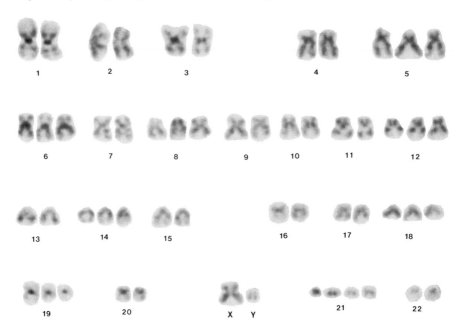

Figure 3. A hyperdiploid karyotype from the bone marrow of a child with ALL. There are 55 chromosomes with additional chromosomes present in each group, except the A group. The chromosomes are highly contracted and of 'fuzzy' ill-defined morphology. No obvious structural abnormalities are present.

to chemotherapy (and possibly radiation therapy) used in the treatment of various malignant states. Secondary leukaemia is an acute problem in its own right as it appears at a much younger age and has a poor response to anti-leukaemic therapy.

7.3 Acute Lymphoblastic Leukaemia

In patients with ALL the metaphases obtained from the leukaemic cells in the bone marrow are characteristically poorly spread, with highly contracted chromosomes of ill-defined morphology. For these reasons progress in the determination of specific chromosome abnormalities in this leukaemia has been slower and less precise than those observed in ANLL. However, certain correlations have emerged between karyotype and prognosis, indicating that karyotype is an independent risk factor in ALL. These findings are described in detail in the Third International Workshop on Chromosomes in Leukaemia (25) and a review article by Secker Walker (26). Briefly, a correlation between chromosome number and patient prognosis is described. For example, children with a hyperdiploid karyotype of greater than 50 chromosomes have the best prognosis (*Figure 3*). These children progress better than those hyperdiploid cases with between 47 and 49 chromosomes. It has been suggested that this group consists of a higher percentage of patients with structural chromosome rearrangements in addition to their numerical abnormalities. Structural rearrangements, in particular translocations, are correlated with a poor prognosis, irrespective of the chromosome number involved. Certain specific translocations have been described in association with reduced patient survival. These include the Ph[1] translocation [t(9;22)(q34;q11)]; t(4;11)(q21;q23); t(8;14)(q24;q32) usually found in those cases with B cell ALL; translocations involving

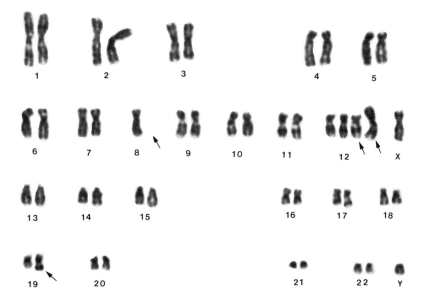

Figure 4. Karyotype from the stimulated B lymphocytes of a patient with CLL, at an advanced stage. In addition to the trisomy 12 other chromosome abnormalities are present, as indicated by the arrows. Note that one of these involves a fourth chromosome 12.

chromosome 14 other than t(8;14) and deletions of chromosome 6. Children with a normal karyotype and those in the hypodiploid group, with a chromosome number between 42 and 45, have an intermediate prognosis, similar to the 47 to 49 hyperdiploid group.

Among the adults with ALL a normal karyotype is associated with the best prognosis, with an intermediate prognosis shown by adult patients in the hyper- and hypodiploid categories. Hypodiploid children and adults with a chromosome number close to haploid have a very poor prognosis associated with a very aggressive form of the disease.

7.4 Chronic Lymphocytic Leukaemia

Trisomy of chromosome 12 is the characteristic abnormality observed in the leukaemic B cells of patients with CLL. This is present in approximately 50% of cases. Patients with trisomy 12 show a poorer prognosis and more rapid disease progression than those cases with a normal karyotype (14). More recently, the group of CLL patients with chromosome anomalies have been subdivided into those with trisomy 12 alone and those in which trisomy 12 is found in addition to other chromosome abnormalities (*Figure 4*). The former group, in association with patients of normal karyotype, belong to the early stages of the disease (Rai Stages I and II), whereas those in the latter group, with additional chromosome abnormalities, are found in the advanced clinical stages of CLL (Rai classification III and IV) (27), coincident with poorer prognosis and shorter survival.

7.5 Non-Hodgkin's Lymphoma

NHLs represent a large group of lymphomas of highly variable histology. Interpretation of the literature relating to specific chromosome abnormalities in NHL is difficult due

157

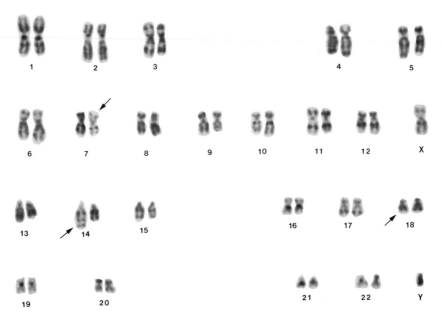

Figure 5. Karyotype from the lymph node cells of a patient with follicular NHL. The characteristic t(14;18) (q32;q21) is present, with a structural rearrangement of chromosome 7.

to the range of histological classifications used between laboratories. Mitelman (22) provides details of all the chromosome abnormalities observed in the various histological subtypes. Irrespective of the classifications used, two major morphological subgroups are present. These are the diffuse and follicular types of lymphoma. One important cytogenetic finding has been described which is diagnostic for the follicular NHLs. This is the translocation t(14;18)(q32;q21) found exclusively in this group of lymphomas (15,28) (*Figure 5*). Other non-random chromosome abnormalities are described by Yunis (15) in association with certain other subtypes using a new international histological classification. These include 6q- in diffuse large cell lymphomas; t(11;14)(q13;q32), deletion of 11q and/or trisomy of chromosome 12 in small cell lymphocytic lymphomas and t(8;14)(q24;q32) in small non-cleaved cell lymphomas. This translocation between chromosomes 8 and 14 is also characteristic of Burkitt's lymphoma. Sub-clonal variation is common in lymphomas resulting in cells with highly complex and variable karyotypes.

7.6 Solid Tumours

A large amount of literature is emerging on the subject of chromosome abnormalities found in a variety of solid tumours. This increase in data has arisen as a result of the improvement of techniques for the handling of tumour material. The presence of a chromosome abnormality is usually an indication of malignancy but the specificity of abnormalities in relation to the tumour type remains, as yet, ambiguous. A list of chromosome anomalies found in solid tumours is presented in Mitelman (22). Much sub-clonal variation may also be observed in the later stages of tumour progression. Therefore, a large number of karyotypes may need to be examined in order to determine the primary changes within the tumours.

158

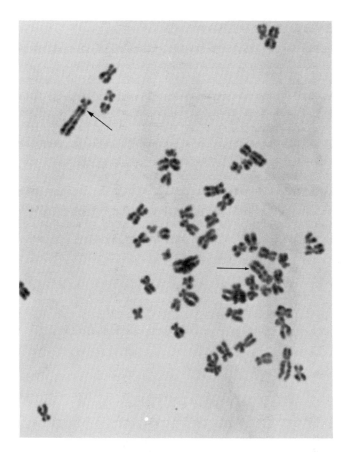

Figure 6. This metaphase from a pleural effusion has a large number of chromosomes, with several chromosome markers (arrows) indicating the presence of a malignancy.

7.7 **Pleural Effusions**

In general, chromosome abnormalities observed in the cells of effusions indicate the presence of a malignancy (*Figure 6*). Specific chromosome abnormalities have been found in a variety of effusions. These findings are reported by Clarke (16). For example, when an effusion is present secondary to a tumour in the gastrointestinal tract the basic clone is often hypodiploid. In effusions from pleural mesotheliomas a range of abnormalities have been found.

(i)   Hypodiploid clones, found in over 50% of cases, usually with around 42 chromosomes. These clones are usually found in diffuse mesotheliomas.

(ii)   Polyploid clones, usually present in mesotheliomas with peritoneal involvement.

(iii) Hyperdiploid clones, found in less than 20% of suspected mesotheliomas.

(iv)   Pseudodiploid clones. These are also rare and are associated with unusual mesotheliomas.

Many structural rearrangements have been found in effusions, but chromosomes 1 and 9 are more frequently involved. Analysis of pleural effusions is labour intensive as often more than 90% of divisions show a normal karyotype.

8. GUIDELINES IN REPORTING CASES

(i)   When reporting the results, bear in mind the definitions laid down by the Fourth International Workshop for Chromosomes in Leukaemia (23) for an abnormal clone: two or more metaphases with identical structural anomalies or extra chromosomes or three metaphases with an identical missing chromosome. One normal cell only is necessary to define the presence of a normal clone.

(ii)   When chromosome banding is not possible but the karyotype is obviously abnormal with, perhaps, numerical abnormalities and/or structural rearrangements resulting in the presence of marker chromosomes, report the findings as any abnormalities are usually an indication of a malignant clone.

(iii)   In cases where random minor structural or numerical anomalies are found it is usual not to report these results. If they occur at a high frequency, report as chromosome instability. In certain of these cases an abnormal clone is subsequently found in repeat samples. Perhaps chromosome instability indicates the future onset of a malignant condition.

9. ACKNOWLEDGEMENTS

I would like to thank Cath Carr and Yvonne Cook for help with the karyotypes and technical advice, and Elaine Mercer for typing the manuscript.

10. REFERENCES

1. Boveri,T. (1914) *Zur Frage der Enstehung maligner Tumoren*, S.Jena: Gustave Fischer.
2. Nowell,P.C. and Hungerford,D.A. (1960) *Science (Wash.)*, **132**, 1497.
3. Neel,B.G., Jhanwar,S.C., Chaganti,R.S. and Hayward,W. (1982) *Proc. Natl. Acad. Sci. USA*, **79**, 7842.
4. Malcolm,S., Barton,P., Murphy,C., Ferguson-Smith,M.A., Bentley,D.L. and Rabbitts,T.H. (1982) *Proc. Natl. Acad. Sci. USA*, **79**, 4956.
5. Erickson,J., Martins,J. and Croce,C.M. (1981) *Nature*, **294**, 173.
6. Heisterkamp,N., Stephenson,J.R., Groffen,J., Hansen,P.F., de Klein,A., Bartram,C.R. and Grosveld, G. (1983) *Nature*, **306**, 239.
7. *Howie Code of Practice for the Prevention of Infection in Clinical Laboratories and Post Mortem Rooms* (1978).
8. Yunis,J.J. (1981) *Hum. Pathol.*, **12**, 540.
9. Hagermeijer,A., Smit,E.M.E. and Bootsma,D. (1979) *Cytogenet. Cell Genet.*, **23**, 208.
10. Weber,L.M. and Garson,O.M. (1983) *Cancer Genet. Cytogenet.*, **8**, 123.
11. Taylor,J.H., Haut,W.F. and Tung,J. (1962) *Proc. Natl. Acad. Sci. USA*, **48**, 190.
12. Yunis,J.J. (1976) *Science (Wash.)*, **191**, 1268.
13. Yunis,J.J., Sawyer,J.R. and Ball,D.W. (1978) *Chromosoma*, **67**, 293.
14. Gahrton,G. and Robert,K.-H. (1982) *Cancer Genet. Cytogenet.*, **6**, 171.
15. Yunis,J.J., Oken,M.M., Theologides,A., Howe,R.B. and Kaplan,M.E. (1984) *Cancer Genet. Cytogenet.*, **13**, 17.
16. Clarke,C. (1984) *Clin. Cytogenet. Bull.*, **1**, 113.
17. Kondo,K. and Sasaki,M. (1981) *Jap. J. Hum. Genet.*, **26**, 255.
18. Nakagome,Y., Yokochi,T., Matsubara,T. and Fukuda,F. (1982) *Cytogenet. Cell Genet.*, **33**, 254.
19. Sumner,A.T., Evans,H.J. and Buckland,R.A. (1971) *Nature New Biol.*, **232**, 31.
20. Seabright,M. (1971) *Lancet*, **ii**, 971.
21. Sandberg,A.A. (1980) *The Chromosomes in Human Cancer and Leukaemia*, published by Elsevier, New York.
22. Mitelman,F. (1983) *Cytogenet. Cell Genet.*, **12**, 1.
23. Fourth International Workshop on Chromosomes in Leukaemia, 1982 (1984) *Cancer Genet. Cytogenet.*, **11**, 251.

24. Sandberg,A.A., Abe,S., Kowalczyk,J.R., Zedgenidze,A., Takeuchi,J. and Kakati,S. (1982) *Cancer Genet. Cytogenet.*, **7**, 95.
25. Third International Workshop on Chromosomes in Leukaemia, 1980 (1981) *Cancer Genet. Cytogenet.*, **4**, 96.
26. Secker-Walker,L.M. (1984) *Cancer Genet. Cytogenet.*, **11**, 233.
27. Han,T., Ozer,H., Sadamori,N., Emrich,L., Gomez,G.A., Henderson,E.S., Bloom,M.L. and Sandberg,A.A. (1984) *New Engl. J. Med.*, **310**, 288.
28. Yunis,J.J., Oken,M.M., Kaplan,M.E., Ensrud,K.M., Howe,R.R. and Theologides,A. (1982) *New Engl. J. Med.*, **307**, 1231.

CHAPTER 6

Meiotic Studies in Man

M.A. HULTÉN, N. SAADALLAH, B.M.N. WALLACE and M.R. CREASY

1. INTRODUCTION

Most investigations of meiosis in man have been concerned with increasing our understanding of the meiotic process itself. The most common application of meiotic studies to clinical cytogenetics relates to the management of carriers of structural chromosomal abnormalities and their risk of producing unbalanced gametes. The experience gained from meiotic investigations of some male carriers helps in assessing the risk for other carriers, but these investigations are time-consuming since they require consecutive banding treatments, and some expertise in interpretation. Furthermore it has so far been impossible to obtain adequate numbers of cells at second metaphase to provide a valid assessment of individual risk. For these reasons the study of the meiotic picture from testicular biopsies of individual male carriers might not be practicable in a busy routine diagnostic cytogenetics service.

Investigations of meiotic cells from testicular biopsies of azoospermic or severely oligospermic males may, on the other hand, help in arriving at a diagnosis, and do not require such detailed analysis. Air-dried preparations from testicular biopsies are adequate for the purpose of detecting a specific meiotic abnormality, and banding is not necessary.

The recent development of the surface-spreading technique, which reveals very clearly pairing at first meiotic division, may also help with precise identification of chromosomal abnormalities whose interpretation from mitotic studies is ambiguous. Analysis of surface-spread preparations with the light microscope, which is all that may be necessary to detect a structural rearrangement such as reciprocal translocation, is a relatively simple procedure.

In this chapter we present the techniques that are used for meiotic investigation in our laboratory, and comment upon the interpretation of the results. Although the difficulty of obtaining oocytes or ovarian biopsies has meant that meiotic investigations have been almost entirely restricted to males, the techniques for the study of female meiosis are included for completeness.

2. REASONS FOR UNDERTAKING MEIOTIC INVESTIGATIONS

2.1 Infertile Males

As infertility is such a common problem, it is not practical to suggest that all men attending an infertility clinic should have testicular biopsies and meiotic investigations

Thorough investigations should be carried out beforehand for other causes of infertility, including mitotic chromosome investigations of infertile men with sperm counts below 20 million/ml. It is mainly in this group that there is an increased incidence of chromosome abnormalities in comparison with the normal population (1).

The most common chromosome abnormality among azoospermic males is an XXY sex chromosome complement associated with Klinefelter's Syndrome. The majority of such men have a complete maturation arrest of spermatogenesis and so meiotic studies would not be informative. Other constitutional chromosome abnormalities associated with azoospermia or oligospermia include structural rearrangements (see Section 2.3), supernumerary chromosomes (see Section 2.4) and XYY males. Meiotic investigations will not help this group with regard to fertility, but would be relevant in assessing the risk of abnormal offspring in the event of a successful pregnancy. In the case of XYY males, the empirical risk of having XYY sons is very low.

For azoospermic men with normal-sized testes and normal mitotic chromosomes, meiotic investigations may help to establish a differential diagnosis. Some such men have a maturation arrest at the first spermatocyte stage, associated with a lower than normal frequency of chiasmata at first metaphase $(1-4)$. This condition may be monogenic with recessive inheritance and these 'oligochiasmate' males, who are described further in Section 3.2, are incurably infertile. In the remaining cases, it may be expedient to continue looking for a cause of the infertility in the hope that it may be treatable.

2.2 **Subfertility**

Couples who have a history of repeated pregnancy wastage, including miscarriage, perinatal death and multiply-malformed children of unknown cause, should have mitotic investigations to detect the presence of a chromosome rearrangement (5). In the absence of any mitotic abnormality, meiotic investigations of the male partner are unlikely to reveal an undetected chromosomal rearrangement.

2.3 **Known Structural Chromosome Abnormalities**

In an ideal world all carriers of structural chromosome abnormalities would be investigated meiotically. However, as it is very difficult to obtain oocytes these studies have been restricted to male carriers. Meiotic investigations from testicular biopsies are of help in assessing fertility. In addition, they reveal how the rearranged chromosomes pair in prophase (6,7) and also the chiasma pattern at first metaphase (8), which is of relevance to the estimation of the relative frequency of the different gametic types (9). It is possible to obtain direct information on first metaphase segregation by karyotyping second metaphases (10) but this is labour intensive and to date no adequate investigations have been performed on carriers of structural rearrangments.

From a practical point of view chromosomal investigations of semen samples would obviously be advantageous in giving a direct estimation of the different types of gametes available but this technique is even more demanding (11,12).

2.4 **Suspected Structural Abnormality**

Most cytogenetic laboratories have at some time experienced difficulty in interpreting the precise nature of a suspected structural chromosome abnormality. In such a case it may be much easier to identify the nature of the rearrangement by meiotic studies: in particular by using the surface-spreading technique (6). From our limited experience in this field, it seems likely that reciprocal translocations are easier to identify than inversions and insertions (unpublished results).

2.5 Apparent *de novo* Chromosome Abnormalities in Offspring

A similar situation arises when a supernumerary chromosome is interpreted as having originated *de novo* because both parents have seemingly normal karyotypes. Consequently, it is suggested that there is no risk of recurrence. Obviously, however, there is a possibility that the apparently *de novo* supernumerary in the offspring is a product generated from a cryptic structural rearrangement in one parent. Investigations of meiotic chromosome pairing in the father may then be helpful, but to date there are no reports in the literature exemplifying this situation.

2.6 Exposure to Environmental Agents

There is scant information on the effect of environmental agents on meiotic chromosomes. Only a few investigations of first meiotic metaphase configurations revealing the induction of reciprocal translocations in human males after exposure to ionising radiation have been published, and their relevance is still uncertain. Nevertheless, it is important to enquire about undue parental exposure in any case of a *de novo* structural chromosome aberration. This type of meiotic investigation, aimed at revealing abnormalities in individual spermatocytes or identifying a meiotic clone, is more exacting than the study of a constitutional chromosome abnormality (13,14).

3. TESTICULAR BIOPSIES

3.1 Obtaining the Material

A testicular biopsy can be taken as an out-patient procedure either under general or local anaesthesia, and only takes about 10 min. An incision of about 1 cm is made in the scrotal skin. The surface of the testis is inspected, and an incision of $2-3$ mm is made in the tunica albuginea. The testicular parenchyma bulges out and is cut with a small pair of scissors. Surgeons are usually reluctant to take biopsies larger than about $1 \times 2 \times 3$ mm, because of the risk of haemorrhage. In our own experience, preparations made from testicular material obtained by needle aspiration are not of as high a quality as those taken by open incision. There is no difference in the quality of the sample whether the biopsy is taken under general or local anaesthesia.

The biopsy must be subdivided for different investigations immediately. Any delay will result in deterioration of the material. Divide the material into three equal parts for testicular histology, surface-spreading and air-dried preparations.

3.2 The Air-drying Technique

3.2.1 *Testicular Cell Suspension and Slide Preparation*

The air-drying technique which we use is based on that of Evans *et al.* (15). These preparations are suitable for examination of premeiotic spermatogonial metaphases, first and second meiotic metaphases. Collect the testicular biopsy in a test tube containing F 10 medium or in 0.075 M KCl.

(i)   Place the tissue in 3 ml of 0.075 M KCl in a small Petri dish.

(ii)   Chop up the seminiferous tubules with fine scissors.

(iii)   Squeeze out the contents of individual tubular segments with a pair of fine forceps, holding the tubule with a needle.

(iv)   Transfer the cell suspension and remains of tubules to a 10 ml test tube and agitate with a pipette to flush out the remainder of the cells from the tubules.

(v)   Leave to stand for 2−3 min. Transfer the supernatant to 2 mini glass test tubes avoiding transferral of tubular remnants.

(vi)   Centrifuge the tubes at 1000 r.p.m. for 8 min, discharge the supernatant and resuspend the cells in 2−3 ml of fixative (30:10:1 methanol:glacial acetic acid: chloroform). Initially, add the fix 1 drop at a time rolling it along the side of the tube. Flick the tube gently while adding the fix. Allow to stand for 5 min.

(vii)   Centrifuge at 1100 r.p.m. for 8 min, discard the supernatant and resuspend the cells in the same way as in step (vi). Allow to stand for 10 min.

(viii) Centrifuge at 1100 r.p.m. for 10 min, discard the supernatant and resuspend the cells in a smaller amount of fix without chloroform.

(ix)   Drop 1−2 drops of the final suspension onto a cooled, clean glass slide and allow to dry. Check the fix of the cells on the first slide. If this is not adequate refix. Prepare as many slides as possible.

(x)   The slides can be stored in 3:1 methanol:glacial acetic acid until required. Screen the slides with a ×10 phase contrast objective and record the coordinates of the cells at different stages of meiosis. The slides may then be stained either by a simple staining technique or sequentially.

3.2.2 *The Quadruple Staining Technique*

This technique was developed by adding distamycin A plus 4,6-diamino-2-phenyl-indole (DAPI) fluorescent staining (16) to the previously described triple staining technique (17) with some modifications (10). Slides are sequentially stained by quinacrine mustard, distamycin A-DAPI, cresyl fast violet [or lacto-propionic orcein (LPO)], and finally C-banded. This order is essential and should not be altered. Slight modifications of the individual techniques had to be introduced to make this complex sequential staining regime work. After each step, photographs should be taken of each suitable cell, so that four photographs, representing the four stages of the staining will be obtained of the same cell.

(i) *Quinacrine mustard fluorescence (Q-banding).*

The staining procedure is as follows.

(1)   Rehydrate the cells with:
        70% ethanol for 5 min;
        50% ethanol for 5 min;
        distilled water for 5 min;
        0.1 M citric acid/0.2 M sodium phosphate buffer (pH 7.4) for 5 min.

(2)   Stain the cells with 50 mg/ml of quinacrine mustard solution for 15 min.

(3)   Rinse the cells in buffer (pH 7.4) for 10 min, mount in buffer and seal the cover-slip with rubber solution (black holdtite rubber adhesive).

(4)   Photograph the cells under a high power objective (e.g. ×63 neofluar objective) with a BG12 filter, using Recordak copy film. The cells should be exposed as little as possible to fluorescence, otherwise subsequent staining procedures may be difficult to perform.

(5)   Wash off the coverslip with tap water and leave the slides to wash in running water for about 30 min.

(ii) *Distamycin A plus DAPI (DA-DAPI) fluorescence.*

(1)   Flood the slides with 0.05 mg/ml distamycin A hydrochloride in McIlvaine's buffer (pH 7), cover with a coverslip and leave for 15 min.

(2)   Remove the coverslip and rinse the slides briefly with McIlvaine's buffer.

(3)   Flood the slides with 0.02 mg/ml DAPI in distilled water, cover with a coverslip and leave in the dark for 15 min.

(4)   Rinse the slides briefly with McIlvaine's buffer.

(5)   Mount the slides in McIlvaine's buffer, remove any excess buffer by gently blotting and seal the slides with rubber solution.

The stained preparations may fade when first examined, but this usually stabilises after a day or so stored in the freezer.

Re-photograph the cells with the appropriate filter (UG1 and UG3).

(iii) *Cresyl fast violet (or lacto-propionic orcein).*

After DA-DAPI fluorescent staining, rinse the slides well with distilled water and stain as follows.

(1)   When dry, stain the slides with a filtered solution of cresyl fast violet (or LPO) for 20−40 min, rinse with distilled water and air-dry.

(2)   Re-photograph the cells as before without a coverslip.

(3   Remove the immersion oil from the slides by treating with Euparal essence and then wash in methanol and dry. This procedure also destains the slides.

(iv) *C-banding.*

(1)   Place dry slides in 1 N HCl for 2−3 min and rinse in distilled water.

(2)   Place the slides in a freshly-prepared, saturated barium hydroxide solution for 3 min at 60°C. The timing at this stage is critical. Rinse the slides well with distilled water.

(3)   Place the slides in 2 × SSC (0.03 g/l NaCl + 0.003 g/l sodium citrate) at 60°C for 5−10 min.

(4)   Rinse the slides in distilled water and stain in 1:4 aqueous Giemsa for 20−30 min.

(5)   Rinse the slides in distilled water, dry and mount in DPX mountant.

(6)   Re-photograph the cells under a high power phase contrast objective using Recordak copy film.

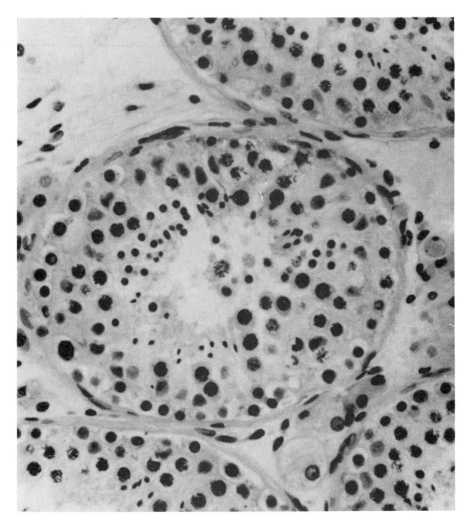

Figure 1. Histological cross-section of a testicular tubule. Spermatogenesis proceeds from the periphery of the tubule towards the centre. Some spermatozoa are seen near the centre of the tubule.

3.2.3 *Analysis*

The photographic negatives from each cell can be mounted as transparencies and labelled. Suitable cells (well-spread and adequately stained) may then be projected to obtain a magnification of about ×2500.

Bivalents can be traced from the cresyl fast violet (or LPO) negatives, and the position of the centromeres superimposed from the C-banding negatives.

3.2.4 *Interpretation of Air-dried Testicular Preparations*

Chromosome preparations are made from a suspension of the cells occupying the testicular tubule (*Figure 1*). Therefore they contain a mixture of interstitial cells (Sertoli cells and Leydig cells) and the cells of the germinal epithelium proper (pre-meiotic spermatogonial cells, first and second spermatocytes, spermatids and spermatozoa). All meiotic stages can be investigated from a single biopsy using the air-drying technique.

As this is a direct preparation, the cell divisions obtained are those occurring spontaneously at the time the biopsy was taken. Only a few cells in first and second metaphase are present on each slide. The following stages of mitosis and meiosis may be observed in cells obtained by the air-drying method.

(i) *Spermatogonial metaphases.*

The mitotic cell divisions observed are spermatogonial. Pre-meiotic spermatogonial metaphases are normally less frequent than the meiotic stages, diakinesis/metaphase I and metaphase II. The majority are diploid but a small proportion is tetraploid. The morphology of the chromosomes may vary from an appearance similar to that of the colchicine pre-treated metaphase of a lymphocyte culture, to highly contracted chromosomes with extended secondary constrictions of chromosomes 1, 9 and 16 and despiralisation of the heterochromatic segment of the long arm of the Y (*Figure 2a* and *b*). Except for these regions the Q- and C-banding pattern appears the same as in somatic cells. Nevertheless karyotyping is difficult due to the contraction of the individual chromosomes in a large proportion of cells and even counting can present a problem when the secondary constrictions are very extended and hardly visible.

(ii) *Meiotic stages.*

(a) *Pachytene.* In studies using air-dried preparations, comparatively little attention has been paid to first meiotic pairing and there are hardly any data on the initial leptotene and zygotene stages. By definition, the homologous chromosomes are fully paired at the later pachytene stage. The X and Y chromosomes are condensed into the so-called sex vesicle, which is often located at the periphery of the nucleus.

When using the standard hypotonic pre-treatment optimal for first and second metaphases, pachytene nuclei are not sufficiently well spread to allow identification of the individual chromosomal pairs (*Figure 3*). This requires a longer hypotonic treatment, which has the disadvantage that the later stages are destroyed. The autosomal bivalents are long and slender, which is probably the main reason why they are difficult to spread adequately without overlapping. When using a chromosome stain such as methylene blue, Giemsa or LPO the bivalents at late pachytene may show a spontaneous banding pattern along the length of the chromosome arms. These bands are referred to as chromomeres. The sequence of the chromomeres characteristic of the pachytene stage is very similar to the mitotic G-banding pattern (18).

In our own experience, Q-staining does not help in the identification of the pachytene bivalents, which show a uniformly bright fluorescence. Crossing-over is believed to take place during the late pachytene stage, but the cross-over sites cannot be identified in air-dried preparations.

(b) *Diplotene.* The following stage, when the homologous chromosomes start to separate, is called diplotene. This stage is rarely seen in air-dried preparations and no detailed investigations of diplotene have been undertaken in the human male. It is presumed that the lack of spermatocytes at diplotene is due to the short duration of this stage, which is in sharp contrast to the situation in the human female (Section 5.5).

(c) *Diakinesis/first metaphase.* Most investigations using traditional air-dried preparations have concentrated on chromosomal behaviour at diakinesis/first metaphase (see 2,19−21). These stages cannot be differentiated as the spindle is destroyed by the hypotonic pre-treatment.

The bivalents have undergone a continuous contraction since pachytene/diplotene.

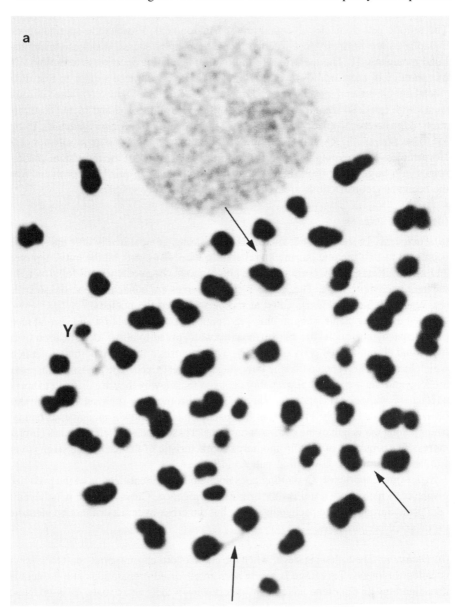

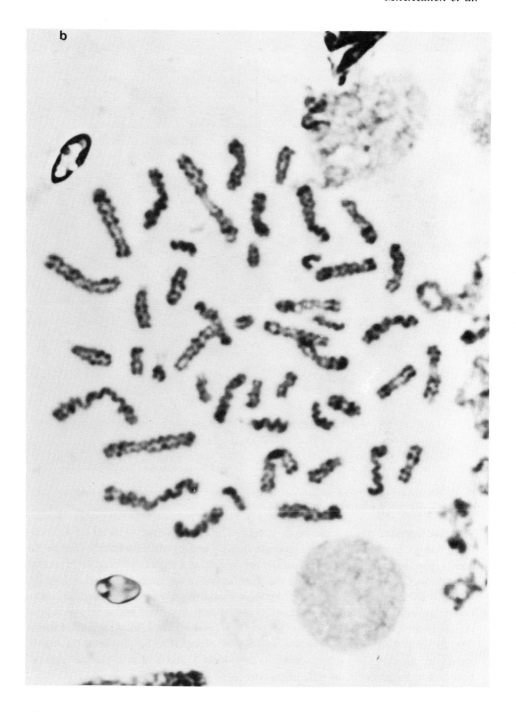

Figure 2. a and **b**. Examples of spermatogonial metaphases from air-dried preparations. Note the contracted chromosomes and the extended secondary constrictions (arrows) as well as the extended heterochromatic part of the Y chromosome in **(a)**.

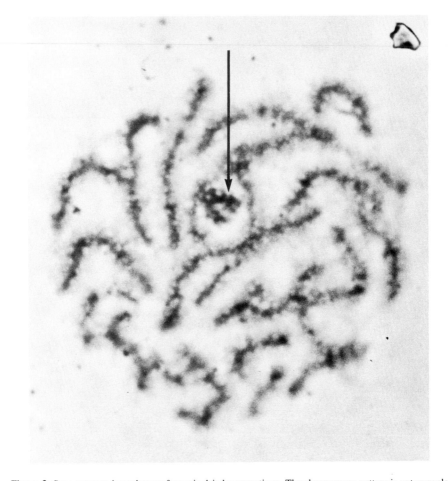

Figure 3. Spermatocyte in pachytene from air-dried preparations. The chromomere pattern is not very clear in this cell. The XY is condensed (arrow).

They are therefore much more easily spread. The only bivalents which may be identified using a plain staining technique are the XY and bivalent number 9 (*Figure 4a and b*). The Q-banding pattern is basically the same as in a mitotic lymphocyte of similar contraction. C-staining is required for the identification of the centromeres (see *Figure 7*).

The X and Y are either associated by their short arm into an XY bivalent or, more rarely, are separate. The XY association is probably chiasmate and a recombination nodule is often seen in this segment (Section 3.3.7 and *Figure 11b*). A chiasma formed near the end of the chromosomes would not be expected to be visible in the highly contracted bivalent. The relevance of the univalents X and Y at this stage to segregation at first anaphase is, as yet, unknown.

The morphology of the autosomal bivalents is dictated by the number and location of the chiasmata. There is a stereotyped pattern of distribution of chiasmata along the length of the bivalents with little variation between individual males, although some atypical males exist (22). All autosomal chromosome arms, with the exception of the

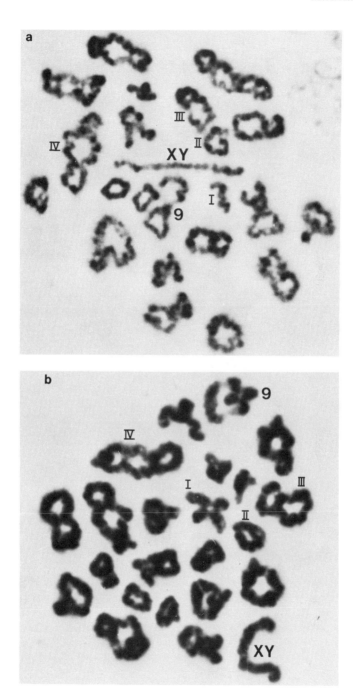

Figure 4. a and **b.** Spermatocytes in diakinesis/first metaphase from air-dried preparations from normal male. Note the XY bivalents which are the only long rods. There is no visible chiasma between the X and Y. Chromosome 9 can be identified because of the extended secondary constrictions. Bivalents with one, two, three or four chiasmata are indicated with Roman numerals (I, II, III, IV).

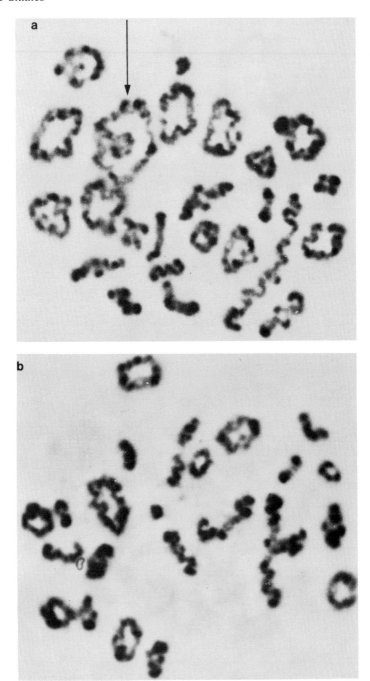

Figure 5. a and **b**. Spermatocytes at diakinesis/first metaphase from air-dried preparations of an infertile man with azoospermia showing reduced chiasma frequency. Note increasing number of rods and some exceptionally long rods having a single chiasma compared with *Figure 4*. The XY bivalent cannot be identified. No bivalent has more than two chiasmata, except for the abnormal structure in **(a)** (arrowed) which is a multivalent probably due to non-homologous pairing.

short arms of the acrocentric chromosomes and number 18, regularly have at least one chiasma (23). The number of chiasmata generally increases with chromosome length. Shorter chromosomes with a single chiasma will form a cross or a rod, depending on whether the chiasma is located in the middle or towards the end. Bivalents with two chiasmata will appear as rings, those with three will give rise to a figure of eight and higher numbers will lead to more complex configurations (*Figure 4a* and *b*). The mean number of autosomal chiasmata in control subjects is 50 and we would at present classify mean total autosomal chiasma frequencies as being abnormal if they were below 44 or above 56, these values being three standard deviations from the mean (23).

A drastic reduction in chiasma frequency may be seen in some infertile males who have azoospermia or severe oligospermia (1 −4) (*Figure 5a* and *b*). It is possible to identify this oligochiasmatic condition without having to apply any more complicated staining techniques than orcein or plain Giemsa. Some bivalents may lack chiasmata altogether and appear as univalents. There is also an increased number of rods and rings in comparison with the more complex configurations normally seen.

Translocations will generally give rise to the expected multivalent (i.e. a trivalent in a Robertsonian and a quadrivalent in a reciprocal translocation). An analysis of the chiasma pattern of the translocation may help to predict first anaphase segregation and chromosome imbalance. The details of this are quite complicated (9), and only some general comments will be made here. If in a reciprocal translocation a chiasma is formed distally in each of the four arms of the pachytene cross (Section 3.3.7, *Figure 13*) a simple ring quadrivalent will be seen at diakinesis/first metaphase (*Figure 6a*). This is the most harmonious type of translocation configuration, and may allow 2:2 segregation at first anaphase to take place in a way similar to the normal with homologous centromeres separating. By this 'alternate' segregation only normal and balanced translocation products result. On the other hand, the same type of 2:2 segregation will lead to 2/4 chromosomally unbalanced secondary spermatocytes if an additional chiasma is formed within either of the segments between the breakpoint and the centromere; the so called interstitial segment. Therefore, from these points of view a low risk is expected if the translocated segments are large enough to allow regular chiasma formation leading to a ring quadrivalent, and the interstitial segments are so short that chiasmata are not often formed here. The risk for individual translocation carriers is increased proportionally to the chance of chiasma formation within the interstitial segment, and additional segregation problems are likely if there is a failure of chiasma formation in one arm of the pachytene cross leading to a chain at diakinesis/first metaphase (*Figure 6b*).

Two further complications may be noted. Firstly, chiasmata are normally clustered in some preferential segments (24,25). Secondly, it is not possible to predict the chiasma pattern for individual translocations in any detail, from the normal situation. There are several examples of a change both within the translocation quadrivalent itself (8,26) and on other chromosomes (8). It may be added that some translocations, particularly those where chain quadrivalents are predominant, may render male carriers subfertile or infertile (Section 3.3.7, *Figure 12*). Female carriers may have an increased risk of chromosomally unbalanced offspring for reasons which are as yet unknown. It should be added that, by comparison with translocation heterozygotes, there have been few investigations of inversion carriers (27−31). Modern staining techniques were not

175

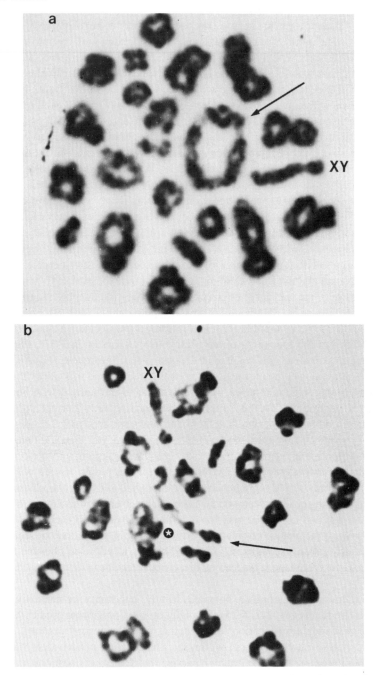

Figure 6. Spermatocyte at diakinesis/first metaphase from air-dried preparations of translocation carriers. **(a)** 46,XY, t(9;10)(p22;q24). Twenty two structures are easily counted, 21 bivalents and a ring quadrivalent, which is easily identified by its large size (arrow). Due to some over-staining it is difficult to identify the chiasmata on some of the normal bivalents. **(b)** 46,XY,t(16q;18q). The quadrivalent appears as a long rod, having only three chiasmata (arrow). It is more difficult to count the number of structures in **(b)** because one small bivalent is overlying a larger one (*).

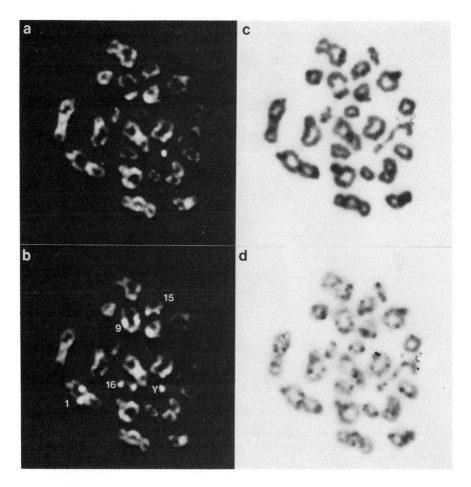

Figure 7. Spermatocyte at diakinesis/first metaphase from air-dried preparations consecutively stained by: (a) quinacrine mustard; (b) distamycin A-DAPI; (c) cresyl fast violet; (d) C-banding.

used, which made it difficult to identify the inversion bivalent and there is no information on the chiasma formation. Such data would be necessary to estimate the frequency of chromosomally unbalanced gametes.

(d) *First anaphase.* To date there have been no investigations of first anaphase. The reason for the lack of studies of anaphase is that this stage is not often seen in meiotic preparations from a testicular biopsy, and there is no specific technique available for the study of first anaphase in the human male.

(e) *Second metaphase.* The following stage is second metaphase, which is difficult to analyse because the chromosomes are contracted and spiralised with splaying of the chromatids. The secondary constrictions of chromosome 1 and especially chromosome 9 may be greatly extended. In addition some of the chromosomes may show separation of the centromeres (*Figure 8*). These problems have meant that only a few adequate

177

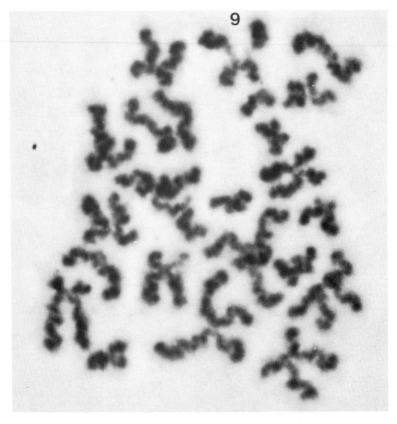

Figure 8. Spermatocyte at second metaphase from air-dried preparation. Chromosomes are spiralised and show splaying of the chromatids. In this cell there is only little separation of the chromatids at the centromere region which makes counting easier. Only chromosome 9 can be positively identified.

karyotypes of second metaphases from either normal males or translocation carriers have been produced. The triple staining procedure is, however, adequate for counting the number of chromosomes present and in many cases it is also possible to karyotype the entire cell (*Figure 9*) (10).

3.3 The Surface-spreading Technique

3.3.1 *Practising the Technique*

The surface-spreading technique demonstrates the synaptonemal complexes (SCs), and is a modification of the techniques of Moses (32), Fletcher (33) and Solari (34). A modification has also been presented by Navarro *et al.* (35).

It is helpful to practise the technique on a laboratory animal before starting on human tissue, because of the difficulty of obtaining this material. The mouse is a useful experimental organism as it is easily available, and the technique for preparing mouse spreads is the same as that used for human testicular material.

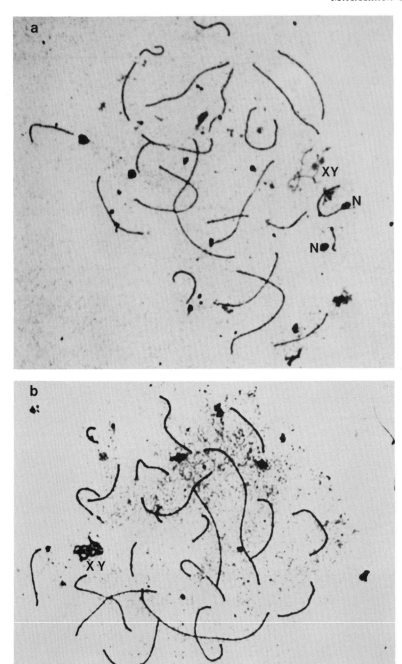

Figure 10. Silver-stained surface-spread spermatocytes at pachytene seen by light microscopy. **(a)** Note the XY bivalent paired by the short arm and the 22 autosomal bivalents. The darkly stained nucleolar organising regions (NOR) help in the identification of the acrocentric bivalents (arrow). **(b)** Later stages of pachytene where the XY is more condensed.

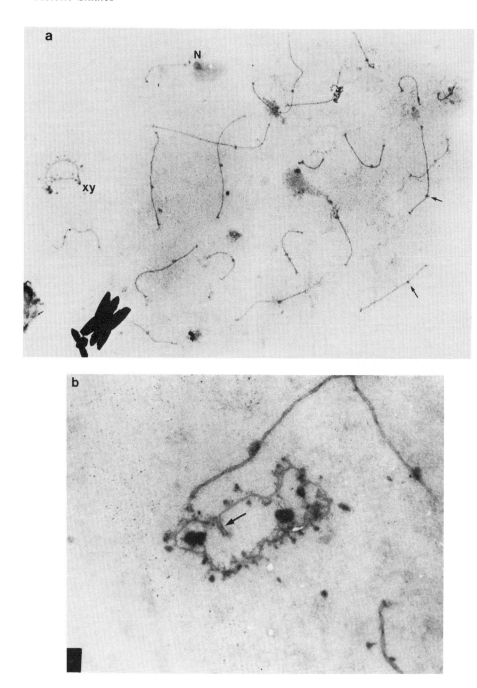

Figure 11. (a) Electron micrograph of pachytene spermatocyte in which all 23 bivalents can be traced along their entire lengths. Two centromeres are arrowed, and one NOR indicated (N). Note the characteristic morphology of the unpaired region of the XY bivalent. **(b)** Enlarged XY bivalent from another cell showing the pairing segment with a recombination nodule (arrow).

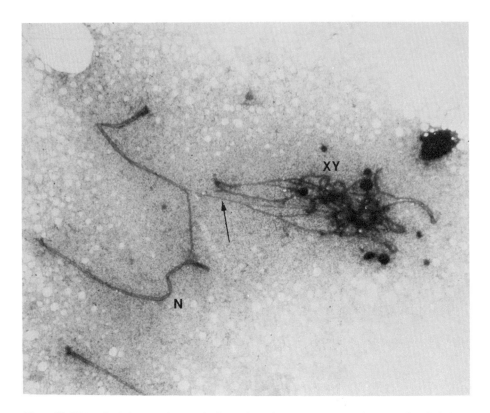

Figure 12. PTA-stained electron micrograph of part of a pachytene spermatocyte from a carrier of a Robertsonian translocation t(13q;14q). The translocation chromosomes form a trivalent. One chromosome is paired normally (N), while the other is paired at its distal end but the proximal part is trapped within the XY complex (arrow).

structures. Zygotene is the following stage when the pairing of homologous chromosomes starts, by pachytene they are fully paired and at diplotene they separate. As already mentioned, investigation of SCs by light microscopy is likely to be adequate for diagnostic purposes. EM investigations will reveal more detail of the complexes, but is more time-consuming. Neither staining technique reveals a clear banding pattern along the length of the bivalents and in our experience no other staining techniques are successful in this respect. The centromeres can sometimes be seen as diffuse patches of chromatin and may be more readily identified by EM. The precise identification of the individual autosomal bivalents presents difficulties by either light or electron microscopy (*Figures 10a* and *b* and *11a* and *b*).

The XY pair has a typical morphology with synapsis initiated at the ends of the short arms (1,34,36). The morphology varies during the different substages of pachytene. Acrocentric chromosomes may be identified by their nucleolar organising regions and there may be a delay in synapsis of the heterochromatic blocks of chromosomes 1, 9 and 16 (6).

The synapsis between homologues generally starts at their telomeres and proceeds towards the centromere. There is normally a high fidelity of homologous synapsis and

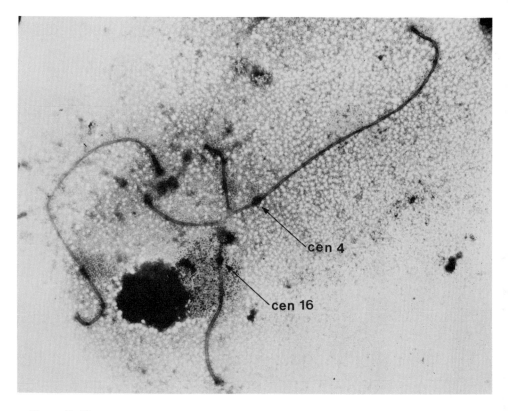

Figure 13. Electron micrograph of a surface-spread quadrivalent from a pachytene spermatocyte from a carrier of t(4;16)(p14;q11.2).

any initial non-homologous association at zygotene is generally resolved at pachytene (*Figure 10a* and *b*). This is the main reason why an analysis of the pairing structures, the SCs, at pachytene may aid in the diagnosis of structural chromosome aberrations.

The limited experience with this technique indicates that a Robertsonian translocation will form a trivalent (7) (*Figure 12*), a reciprocal translocation a quadrivalent (*Figure 13*) and a more complex three-chromosome translocation gives rise to a hexavalent (*Figure 14*) (6). The only other type of structural aberration which has so far been studied by this technique is pericentric inversion (unpublished results). Our impression is that small inversions may be difficult to identify as a loop may not be formed (*Figure 15a*). Failing loop formation, the inverted segments may remain unpaired and show a morphology similar to that of the differential unpaired segments of the X and Y, or non-homologous synapsis may occur (*Figure 15b*). Large inversions may more often demonstrate loops (*Figure 16*). Normally there is some association between one or other autosomal SC and the XY bivalent in about 10% of cells. A specific increased association between the XY and a structurally rearranged SC may occur (7) (*Figure 12*). It has been suggested that such association might interfere with the inactivation of the X chromosome, which is believed to be the prerequisite for maturation of spermatocytes (37). This could explain the impaired spermatogenesis and reduced fertility seen in some

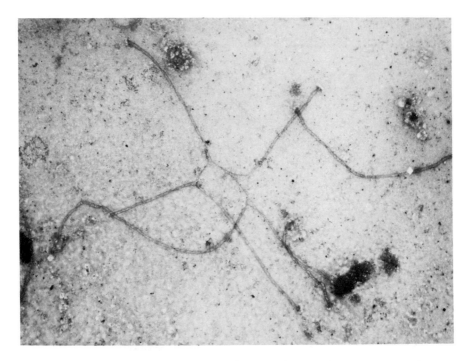

Figure 14. Electron micrograph of a surface-spread hexavalent from a pachytene spermatocyte from a carrier of a complex translocation t(2;9;4)(p13;p12;q24).

male carriers. It seems likely that reduced pairing in itself may also play a part and cells with lack of adequate pairing might be selected against and fail to reach first metaphase (38).

4. OTHER TECHNIQUES FOR THE STUDY OF MEIOSIS IN THE MALE

As many men are reluctant to give a testicular biopsy, the investigation of semen samples is an attractive alternative. There has been some indication in the literature (39) that, at least in oligospermic men, it may be possible to obtain a few cells in meiosis (which have been shed from the germinal epithelium into the lumen of the tubule) from a semen sample. In our own experience, it is then more difficult to obtain good differentiation of the chromosomes, which are usually of poorer quality than those obtained from a testicular biopsy.

The nuclei of spermatozoa decondense and the chromosomes can be resolved after penetration of zona-free, isolated hamster oocytes. This technique is however time-consuming and difficult to set up. Since it was introduced by Rudak *et al.* (40) it has been attempted in many laboratories but only a few have been successful in obtaining a sufficient number of male pronuclei with adequate chromosome morphology to allow karyotyping. Readers are referred to the articles by Martin (11) and Brandriff *et al.* (12) giving details of this exciting technique. It would obviously be valuable if alternative, and simpler techniques could be developed.

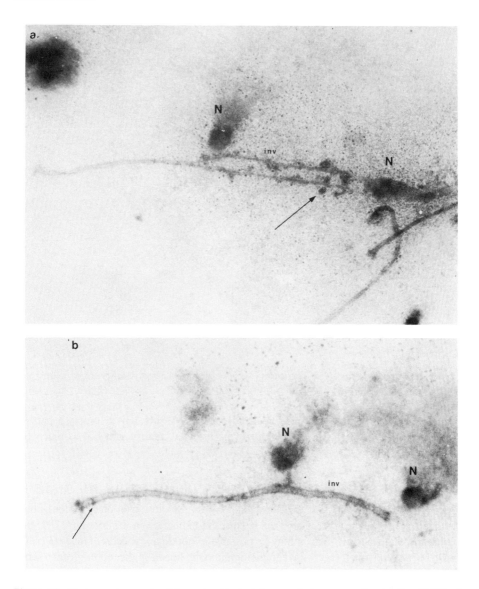

Figure 15. Electron micrograph of inversion bivalent from pachytene spermatocyte of a 46,XY of inv(13)(p12q14 carrier). **(a)** The inverted segment is unpaired (arrow) and shows similar morphology to the unpaired differential axes of the XY bivalent in a normal cell. Two NOR regions (N) can be seen because one breakpoint is within the NOR. **(b)** The inverted segment shows non-homologous pairing. The arrow points to a recombination nodule, which is presumed to be the site of crossing-over.

5. FETAL OVARIES

5.1 Obtaining the Material

Fetal ovaries are usually obtained after termination of pregnancy at about 20 weeks of gestation after antenatal diagnosis of a fetal chromosome abnormality. The ovaries

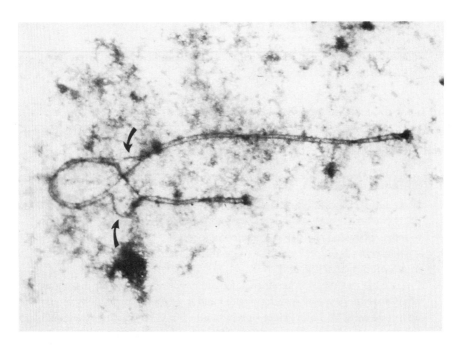

Figure 16. Electron micrograph from a pachytene spermatocyte of a 46,XY of inv(2)(p13q35 carrier). The bivalent shows the classical inversion loop. The regions at the base of the loop are unpaired (arrows).

should be removed as soon as possible and placed immediately in a suitable tissue culture medium, such as Ham's F10, for transport. Processing should take place as quickly as possible after removal of the ovaries.

5.2 The Air-drying Technique for Ovaries

This technique, which can be used for mammalian ovaries from any source including human fetuses, was initially developed by Luciani *et al.* (41). Ovaries may be collected in a balanced salt solution, such as Hank's BSS, or tissue culture medium, such as Ham's F10, and then fixed. It is preferable to fix the ovaries directly as described below. Once fixed, the ovaries can be stored more or less indefinitely before processing. A hypotonic treatment can be included but yields no more detail and the nucleoli are destroyed.

(i)   Immediately after removal place the ovary in 3:1 methanol:glacial acetic acid.
(ii)   Finely mince the ovary with a scalpel.
(iii)   Transfer the fragments of ovary to a conical-bottom test tube.
(iv)   Remove the excess of fixative but do not discard.
(v)   Add about 10 drops of freshly prepared 45% acetic acid, thoroughly mix and leave for 3−4 min.
(vi)   Drop the acetic acid, which now contains the germinal cells in suspension, onto cold, wet slides and leave to dry.
(vii)   Centrifuge the fixative which was removed during stage (iv) at 800 r.p.m. for 8 min.
(viii) Remove the supernatant and resuspend the pellet in a few drops of 45% glacial acetic acid.

(ix)   After thorough mixing, drop the suspension onto cool, wet slides and leave to dry, or

(x)   alternatively, a few drops of methanol can be added to the suspension which is then dropped onto dry slides.

5.3 Preparation of Surface-spreads for Light Microscopy

This technique is a modification of that used for spermatocytes (32,33,34) as described by Wallace and Hultén (42).

5.3.1 *Preparation of Materials*

(i)   Make a 0.03% solution of 'Joy' in distilled, deionised water (43) and adjust to pH 8.5 with 10 mM sodium tetraborate. It can then be stored at 4°C for 2−3 months, provided the pH is checked before use.

(ii)   Dispense the solution from a syringe through a 0.22 μm pore Millipore filter in a Swinnex attachment.

(iii)   Prepare the fixative, 4% paraformaldehyde in 0.1 M sucrose, by dissolving 3.4 g of sucrose in 70 ml of distilled water, add 4 g of paraformaldehyde and make up to 100 ml with water. Heat slowly to 60−70°C while stirring and add about six drops of 1 N NaOH; continue stirring until the solution is clear. Allow to cool, adjust to pH 8.5 with 0.2 M borate buffer and filter (32). Store and dispense in the same way as the detergent.

(iv)   Clean microscope slides with a non-abrasive cleaning cream (e.g. Jif, Lever Brothers), rinse and dry, to give a hydrophilic surface.

(v)   Freshly dissolved silver nitrate is made for each staining and dispensed through a filter as already described. Make up a 50% dilution in distilled de-ionized water containing 0.04% formaldehyde.

5.3.2 *Preparation of the Slides*

Pre-treating the ovary for 20 min with hypotonic (0.075 M) KCl does not produce a marked improvement. The technique detailed below was found to be the most consistent in providing the greatest number of well-spread nuclei without excessive stretching of the SCs.

(i)   Transfer the ovary and medium to a small Petri dish kept on ice. Remove adherent tissue and squash a small sample of ovary in orcein to confirm the presence of meiotic stages.

(ii)   Tease a piece of ovary about 0.5 mm in diameter in two drops of Joy on a slide, and allow the cells to disperse for 6 min at room temperature. Remove any fragments of ovary.

(iii)   Fix the preparation by adding five drops of paraformaldehyde and leave the slide to dry on a hot-plate at 30°C for at least 2 h or overnight.

(iv)   Rinse the slides in distilled water adjusted to pH 8.5, air-dry and leave for several days before staining.

(v)   Add two drops of silver nitrate solution to each slide, and then cover with a 22 × 50 mm coverslip. Incubate at 55°C for 21 h in a moist chamber.

(vi)   Dip the slides in distilled water so that the coverslips fall off, rinse in a stream of distilled water, air-dry and mount in Euparal Vert.

5.3.3. *Analysis*

Scan the slides at low magnification and photograph selected clear cells using an oil immersion ×100 objective and 35 mm Recordak AHU microfilm. Bivalents can be identified from enlarged photographic prints, referring back to the original preparation to resolve any difficulties. The SC complements can be traced from these prints and measured using a digitiser linked to a computer, if available.

5.4 Preparation of Surface-spreads for Electron Microscopy

A similar method may be used for the preparation of surface-spreads for electron microscopy. However, in this case the cells cannot be teased directly on the slide which is coated with plastic and care must be taken not to dislodge the film while rinsing slides. The spreading solution and fixative are the same as used for light microscopy.

(i)   Coat cleaned slides with 0.5% optilux (Falcon plastic Petri dish) or 0.5% Pioloform (Wacker-Chemie, Munich) in chloroform and either glow-discharge (10 Torr 2 min at 1.3 kV) or treat with 0.1% poly L-lysine for 30−60 min (44). Rinse with distilled water (drying is not necessary) to render the slides hydrophilic.

(ii)   Transfer the ovary and medium to a small Petri dish kept on ice. Remove adherent tissue and squash a small sample of ovary in orcein to confirm the presence of meiotic stages.

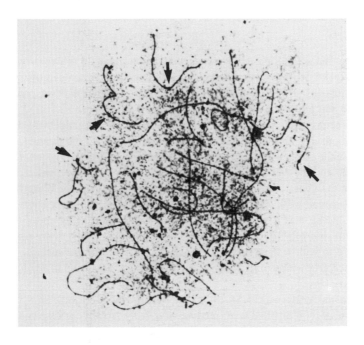

Figure 17. Silver-stained surface-spread, pachytene oocyte from a chromosomally normal 19-week anencephalic fetus. The 23 bivalents are fully paired. Centromeres are indicated by arrows.

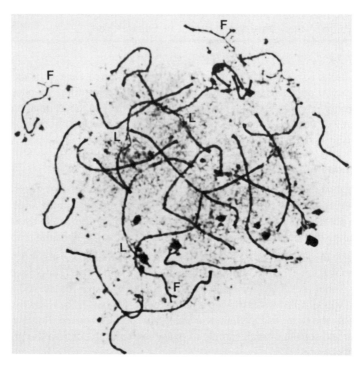

Figure 18. Silver-stained surface-spread, early diplotene oocyte from a chromosomally normal 22-week fetus. Several bivalents are still completely paired but others show desynapsis (arrows) at the telomeres (F) or interstitially (L).

(iii)   Tease out a small piece of ovary about 1 mm diameter in two drops of 0.03% Joy (pH 8.5), in a solid watch glass. Remove the fragments of ovary and dilute the suspension with six drops of Joy. Pipette two drops of suspension onto a slide and allow the cells to disperse for 6 min.

(iv)   Fix the preparation by adding five drops of 4% paraformaldehyde and leave the slides to dry on a hot-plate at 30°C for at least 2 h or overnight.

(v)   Rinse the slides with a gentle jet of distilled water (pH unadjusted) from a wash bottle and air-dry.

(vi)   Scan the slides at low magnification using a phase contrast objective to locate well-spread cells. Place an EM grid (Maxtaform H61) made slightly adhesive by dipping it in a solution of Sellotape in chloroform (2.5 cm of Sellotape to 25 ml chloroform) over suitable well-spread preparations. Score round the film and grid with a diamond marker, and float off on a water surface. Pick up the filmed grid with a thick paper towel or parafilm, and leave to dry in a dust-free atmosphere.

(vii)   Stain the grids with 7.5% uranyl acetate in 50% ethanol for 15 min, wash thoroughly in distilled water and air-dry. Alternatively, grids can be stained in alcoholic PTA (4% PTA diluted 1:3 with 95% ethanol) for 5 min, rinsed in 95% ethanol and air-dried. Note, however, that alcoholic PTA is not suitable for staining Pioloform-coated grids.

192

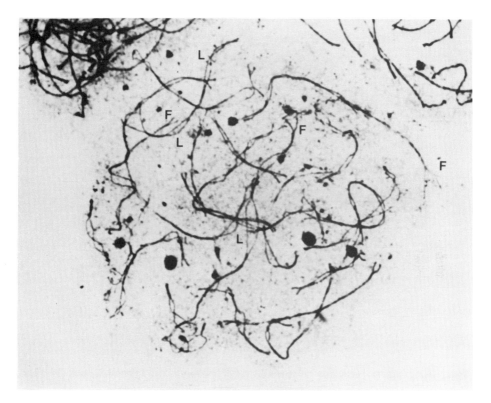

Figure 19. Silver-stained surface-spread, mid-diplotene oocyte from a normal 22-week fetus. The telomeric forks (F) and interstitial loops (L) have increased in size and number.

(viii) Map the position of spreads on the grids at low magnification, using a phase contrast objective, before examination with the electron microscope. Photographs are taken on Agfa Gevaert Scientia 23 D56 or Ilford Electron Microscope film.

Slides may also be stained for light or electron microscopy with silver nitrate (45,46).

5.5 Interpretation of Surface-spread Oocytes

The surface-spread technique is well suited to the study of female meiotic stages from leptotene to early diplotene and, in particular, to pachytene (*Figure 17*). At this stage the chromosomes are twice as long in oocytes as spermatocytes (42) which increases the likelihood of chromosome overlapping resulting in fewer clear spreads. Leptotene is difficult to identify in the light microscope and only a few examples of zygotene have been found, which suggests that pairing occurs quite rapidly. In early zygotene the chromosomes are too thin and entwined to allow analysis.

As in spermatocytes, centromeres are seldom visible in the light microscope but are more frequently resolved in the electron microscope. Nucleoli are usually dispersed by spreading. This hinders karyotyping and only bivalents 1, 21 and 22 can be identified with any certainty at pachytene. Unlike the male, where the sequence of changes in the XY pair can be used to subdivide pachytene, in the female there are no indepen-

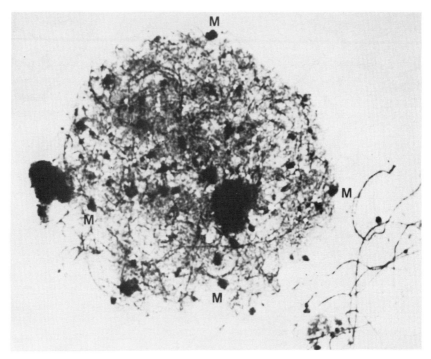

Figure 20. Silver-stained surface-spread, late diplotene diffuse oocyte from a normal 22-week fetus. The bivalents cannot be traced as the SC has partly disintegrated. Numerous micronuclei (M) are present.

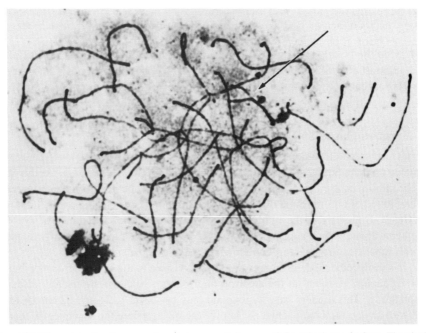

Figure 21. Silver-stained surface-spread, pachytene oocyte from an 18·5-week trisomy 21 fetus. The trivalent with a single fork is arrowed.

dent markers for staging pachytene. For further details regarding the female prophase stages readers are referred to the article by Speed (47).

As in spermatocytes, synapsis of homologues starts at the chromosome ends and proceeds towards the centromeres. This pattern is not, however, universal as a few cases of interstitial initiation have been seen and, in particular, the acrocentrics may have a different sequence with the centromeres and short arms pairing last. At pachytene pairing is complete with the occasional exception of short regions interpreted as the heterochromatic blocks of bivalents 1, 9 and 16.

In contrast to the male, diplotene in the human female is a long stage which may last for many years. Many oocytes reach diplotene at about 22 weeks post-conception and remain in diplotene until just prior to ovulation, when the chromosomes condense and complete the meiotic divisions. Few oocytes reach metaphase, the majority being lost through atresia.

The onset of diplotene is recognised in surface-spreads by desynapsis of the telomeres on some of the bivalents. At this stage pairing must have relaxed along many of the bivalents, as loops of unpaired interstitial regions are usually present along many bivalents of the same nucleus (*Figure 18*). The number and size of these terminal unpaired forks and interstitial loops apparently increases progressively during diplotene (*Figure 19*) until the bivalents can no longer be traced (*Figure 20*).

Structural rearrangements have not been studied in the human female but oocytes trisomic for chromosome 21 form either a bivalent and univalent or a trivalent configuration clearly visible in the light microscope (*Figure 21*). The trivalents are associated either partially or completely along their length to give a variety of forked structures, the majority being single or double forks. The added resolution of the electron microscope shows that the lateral elements of the SC in the trivalents are associated, demonstrating that, contrary to the classical view of chromosome pairing, three homologous chromosomes can be held in register at the same site (48−50).

6. ADULT OVARIES

It is not easy to obtain oocytes for investigation of first and second metaphase and we have no personal experience in our laboratory with these techniques. Some information is given in the recent review article by Hultén *et al.* (14) and further information in the references therein.

7. ACKNOWLEDGEMENTS

We are grateful to Mr D.Matthews for help with the illustrations and Miss C.Robinson for typing the manuscript.

8. REFERENCES

1. Chandley,A.C. (1984) in *Oxford Reviews of Reproductive Biology,* Clarke,J.R. (ed.), Clarendon Press, Oxford, Vol. **6**, 1.
2. Hultén,M.A. and Lindsten,J. (1973) *Adv. Hum. Genet.,* **4**, 237.
3. Chandley,A.C. (1981) in *Oligozoospermia. Recent Progress in Andrology,* Frajese,G. (ed.), Raven Press, New York, p. 247.
4. Micic,M., Micic,S. and Diklic,V. (1982) *Clin. Genet.,* **22**, 266.
5. Tharpel,A.T., Tharpel,S.A. and Bannerman,R.M. (1985) *Br. J. Obstet. Gynaecol.,* **92**, 899.

6. Saadallah,N. and Hultén,M.A. (1985) *Hum. Genet.*, **71**, 312.
7. Rosenmann,A., Warhman,J., Richler,C., Voss,R., Persitz,A. and Goldman,B. (1985) *Cytogenet. Cell. Genet.*, **39**, 19.
8. Laurie,D.A., Palmer,R.W. and Hultén,M.A. (1984) *Hum. Genet.*, **68**, 235.
9. Hamerton,J.L. (1971) *Human Genetics, Vol.* **1**, Academic Press, New York.
10. Laurie,D.A., Firkett,C.L. and Hultén,M.A. (1985) *Ann. Hum. Genet.*, **49**, 23.
11. Martin,R.H. (1983) *Cytogenet. Cell Genet.*, **35**, 252.
12. Brandriff,B., Gordon,L., Ashworth,L., Watchmaker,G., Cartano,A. and Wyrobek,A. (1984) *Hum. Genet.*, **66**, 193.
13. Hultén,M.A., Laurie,D.A., Martin,R.H., Saadallah,N. and Wallace,B.M.N. (1984) in *Proceedings of the American and Swedish Workshop on Individual Susceptibility to Genotoxic Agents in the Human Population,* De Serres,F., Pero,R. and Sheridan,W. (eds.), Plenum Press, New York, p. 441.
14. Hultén,M.A., Wallace,B.M.N., Saadallah,N. and Cockburn,D.J. (1985) in *Aneuploidy.* Etiology and Mechanisms, Dellarco,V.L., Voytek,P.E. and Hollaender,A. (eds.), Plenum Press, New York, p. 75.
15. Evans,E.P., Breckon,G. and Ford,C.E. (1964) *Cytogenetics,* **3**, 289.
16. Schweizer,D., Ambro,P. and Andrie,M. (1978) *Exp. Cell Res.*, **III**, 327.
17. Laurie,D.A., Hultén,M.A. and Jones,G.H. (1981) *Cytogenet. Cell Genet.*, **31**, 153.
18. ISCN (1985) *An International System for Human Cytogenetic Nomenclature* (1985) Birth Defects Original Article Series, **21**, No. 1.
19. Hultén,M.A. and Lindsten,J. (1970) in *Human Population Cytogenetics*, Jacobs,P.A., Price,W.H. and Law,P. (eds), Pfitzer Medical Monographs, Edinburgh University Press, Vol. **5**, p. 24.
20. Chandley,A.C. (1975) in *Modern Trends in Human Genetics*, 2nd edn., Emery,A.E.H. (ed.), Butterworths, London, p. 31.
21. Chandley,A.C., Seuanez,S.H. and Fletcher,J. (1976) *Cytogenet. Cell. Genet.*, **17**, 98.
22. Laurie,D.A. and Hultén,M.A. (1985) *Ann. Hum. Genet.*, **39**, 203.
23. Laurie,D.A. and Hultén,M.A. (1985) *Ann. Hum. Genet.*, **49**, 189.
24. Hultén,M.A. (1974) *Hereditas*, **76**, 55.
25. Hultén,M.A., Luciani,J.M., Morrazinni,M.L. and Kirton,V. (1978) *Cytogenet. Cell Genet.*, **22**, 37.
26. Palmer,R.W. and Hultén,M.A. (1983) *Ann. Hum. Genet.*, **37**, 299.
27. Hauksdottir,H., Halldossen,S., Jensson,O., Mikkelsen,M. and McDermott,A. (1972) *J. Med. Genet.*, **9**, 413.
28. De la Chapelle,A., Schroder,J., Senstrand,K., Fellman,J., Herva,R., Saarni,M., Anttolainen,J., Tallila,L., Husa,L., Tallquist,G., Robson,E.B., Cook,P.J.L. and Sanger,R. (1974) *Am. J. Hum. Genet.*, **26**, 746.
29. Van der Linden,A.G.J.M., Pearson,P.L. and Van de Kamp,J.J.P. (1975) *Cytogenet. Cell Genet.*, **14**, 126.
30. Winsor,E.J.T., Palmer,C.G., Ellis,P.M., Hunter,J.L.P. and Ferguson-Smith,M.A. (1978) *Cytogenet. Cell, Genet.*, **20**, 169.
31. Faed,M.J.W., Lamont,M.A. and Baxter,D. (1982) *J. Med. Genet.*, **19**, 49.
32. Moses,M.J. (1977) *Chromosoma*, **60**, 99.
33. Fletcher,J.M. (1979) *Chromosoma*, **65**, 271.
34. Solari,A.J. (1980) *Chromosoma*, **1**, 315.
35. Navarro,J., Vidal,F., Guitart,M. and Egozcue,J. (1981) *Hum. Genet.*, **59**, 419.
36. Chandley,A.C., Goetz,P., Hargreave,T.B., Joseph,A.M. and Speed,R.M. (1984) *Cytogenet, Cell. Genet.*, **38**, 241.
37. Forejt,J. (1982) in *Genetic Control of Gamete Production and Function*, Crosignani,P.G., Rubin,B.L. and Fraccaro,M. (eds), Academic Press, New York, p. 135.
38. Burgoyne,P.S. and Baker,T.G. (1984) in *Controlling Events in Meiosis*, Evans,C.W. and Dickinson,H.G. (eds), Cambridge, Company of Biologists, p. 349.
39. Templado,C., Marina,S., Coll,M.D., Egozcue,J. (1980) *Hum. Genet.*, **53**, 335.
40. Rudak,E., Jacobs,P.A. and Yanagimachi,R. (1978) *Nature*, **274**, 911.
41. Luciani,J.M., Devictor-Vuillet,M., Gagne,R. and Stahl,A. (1974) *J. Reprod. Fertil.*, **36**, 409.
42. Wallace,B.M.N. and Hultén,M.A. (1985) *Ann. Hum. Genet.*, **49**, 215.
43. Miller,O.L.,Jr. and Bakken,A.H. (1972) *Acta Endocrinol., Suppl.* **168**, 155.
44. Mazia,D., Schatten,G. and Sale,W. (1975) *J. Cell Biol.*, **66**, 198.
45. Howell,W.M. and Black,D.A. (1980) *Experientia*, **36**, 1014.
46. Kodama,Y., Yoshida,M.C. and Sadaki,M. (1980) *Jap. J. Hum. Genet.*, **25**, 229.
47. Speed,R.M. (1985) *Hum. Genet.*, **69**, 69.
48. Wallace,B.M.N. and Hultén,M.A. (1983) *Ann. Hum. Genet.*, **47**, 271.
49. Johanisson,R., Gropp,A., Winking,H., Coerdt,W., Rehder,H. and Schwinger,E. (1983) *Hum. Genet.*, **63**, 132.
50. Speed,R.M. (1984) *Hum. Genet.*, **66**, 176.

CHAPTER 7

Specialist Techniques in Research and Diagnostic Clinical Cytogenetics

S.MALCOLM, J.K.COWELL and B.D.YOUNG

1. INTRODUCTION

The study of the human karyotype has been greatly aided by a number of recent technical advances. Although conventional banding techniques will remain the mainstay of routine cytogenetic analysis, we can anticipate that other approaches will have an increasing impact. For example, flow cytometry of the human karyotype can provide high resolution analysis of certain chromosomal changes and may ultimately form the basis of an automated form of karyotype analysis. Human gene mapping can now be approached by a variety of techniques including flow cytometry, *in situ* hybridisation and somatic cell hybridisation. All of these approaches are now being combined with recombinant DNA technology to provide powerful techniques for the analysis of the human karyotype. In this chapter, we discuss the potential of these experimental techniques in relation to their impact on cytogenetic analysis.

2. IN SITU HYBRIDISATION

2.1 Introduction

In the last 4 years *in situ* hybridisation has become established as a central method of human gene mapping. Its advantage over other methods lies in the excellent regional localisation it provides. Not only does it identify the chromosome on which a gene lies, but it identifies the region to a resolution of about 0.5% of the genome ($1-2 \times 10^7$ bp). This has been of great significance in the study of the role of oncogenes in certain tumours (1,2). There are two fundamental aspects of the technique:

(i)   melting the two DNA strands of the double helix and reannealing them to radioactively labelled probe DNA;
(ii)   identifying the chromosomes after hybridisation and autoradiography.

To a large extent these two aims are contradictory and conditions must be chosen which give a satisfactory compromise between the two. The majority of banding techniques either will not work at all, or work at greatly reduced efficiency on denatured chromosomes. The final choice of banding technique, however, often depends on experience, prejudice and apparatus already available in the laboratory, and most methods can probably be adapted with trial and error.

2.2 **Sensitivity**

Simple calculations, taking into account factors such as specific radioactivity of the DNA probe, length of the gene and autoradiographic efficiency, suggest that localisation of a single copy gene is virtually impossible within a reasonable time (i.e., duration of a grant). However, a number of factors invalidate this pessimistic conclusion. If dextran sulphate is incorporated into the hybridisation buffer, the effective concentration of nucleic acid is increased, and formation of networks between overlapping single strands of radioactive DNA occurs. The entire cluster contributes to the radioactive signal observed over the region of the chromosome to which it hybridises. Secondly, the sequences used for hybridisation need not correspond to coding regions of the gene but may include intervening and flanking sequences providing these contain no repetitive sequences. Any repeated sequences can be removed by sub-cloning, and more than one piece of closely linked DNA can be used in an experiment. Sequences of a total length of 5 kb are ideal for mapping. Sequences down to 1 or 2 kb can be used where necessary and even shorter sequences have been reported. Even increasing the signal as much as possible will not result in many silver grains, and so not only does great care have to be taken to remove any background grains, but it is also necessary to pool data from a number of chromosome spreads to give a statistically significant result.

2.3 **Source of Chromosomes**

Chromosomes produced by phytohaemagglutinin (PHA) stimulated peripheral blood lymphocytes fixed in 3:1 methanol:acetic acid (see Chapter 2) are ideal for *in situ* hybridisation. Chromosomes prepared from carriers of balanced translocations are particularly useful as the breakpoint can be determined on non-hybridised chromosomes, and the position of the grains relative to the breakpoint determined (3). There is an advantage in using more extended chromosomes which can be achieved by using bromodeoxyuridine (BrdU) or methotrexate synchronisation (see Chapter 2) (4). Even greater resolution can be achieved by using meiotic chromosome preparations (5), in which case individual pachytene bivalents are identified by chromomere patterns, thus eliminating the need for banding. Although somewhat less easy to work with, both tumour cells (1,2,6) and fibroblasts carrying translocations (7) can be used.

2.4 **Choice of Isotopes**

Unfortunately, in choice of isotope for radio-labelling there is a conflict between sensitivity and resolution, as the isotopes which give the highest specific activity have disintegrations which are so energetic that they produce tracking in the emulsion. Therefore, ^{32}P, which is the isotope of choice for blotting experiments, is quite unsuitable. ^{3}H gives the best resolution but the lowest specific activity, whereas ^{125}I will increase sensitivity but at the expense of resolution (see, for example, reference 7, Figure 2). ^{35}S-Labelled nucleotides should provide an ideal compromise, but so far they have not been exploited for gene mapping. All these labelled nucleotides are available commercially. Much has been written about the possibility of identifying hybridised probes by cytochemistry or antibody labelling, and this would confer enormous advantages by removing the autoradiography step, and by speeding up the procedure. Both bio-

Table 1. Banding Methods.

A. *Lipsol Banding*

1.   Immerse the slides for $3-10$ sec in a 0.5% solution of Lipsol made up in phosphate-buffered saline (PBS).
2.   Wash off Lipsol thoroughly using PBS.
3.   Stain the slides in 5% Giemsa for 5 min.

B. *Wright's Staining*

The chromosomes are banded through the photographic emulsion.
1.   Wright's stain is diluted 1:3 with 0.06 M phosphate buffer (pH 6.8) for $8-10$ min.

The bands become more evident if the slides are de-stained and re-stained as follows.
2.   Immerse the slides in 95% methanol for 2 min.
3.   Immerse the slides in chloroform for 15 sec.
4.   Immerse the slides in 95% ethanol containing 1% HCl for 30 sec and 100% methanol for 2 min.
5.   Re-stain the slides for $6-8$ min.

C. *Acridine Orange R-Banding*

Chromosomes are banded after hybridisation.
1.   Incubate the slides in 0.07 M phosphate buffer (pH 5.2) at 87°C for 5 min. Old slides will require a longer treatment.
2.   Stain in 0.01% acridine orange solution in 0.07 M phosphate buffer (pH 6.5) for 5 min.
3.   Wash the slides for $2-8$ min in phosphate buffer (pH 6.5), cover the slides with this buffer and place a coverslip on top.

D. *Quinacrine Mustard Q-Banding*

1.   After hybridisation, place the slides in 0.005% quinacrine mustard solution made up in 0.1 M citric acid, 0.2 M $NaHPO_4$ (pH 5.4) for 20 min.
2.   Rinse the slides 10 times in the buffer and then soak for a further 10 min in the dark.

---

tinylated (8) and 2-acetylaminofluorene (9) modified probes have been used to locate repetitive sequences, but the sensitivity of the technique does not yet extend to single copy sequences.

2.5 Banding Techniques

2.5.1 *Pre-banding versus Post-banding*

Two methods are available for identification of chromosomes after labelling:

(i)   those in which cells are banded and photographed before denaturation and then the same cells relocated at the end of the procedure (pre-banding);
(ii)   those in which no banding is attempted until the whole experiment is finished and the autoradiographs developed (post-banding).

The most important features to be considered are that pre-banding removes any possible observer bias in selecting cells for analysis, whereas post-banding allows many more cells to be analysed, but may result in complete failure of the experiment if the chromosomes do not band.

(i) *Pre-banding*. The most widely used pre-banding method employs a dilute solution of Lipsol (LIP Equipment and Services Ltd). A full protocol is given in *Table 1*. The time of immersion in Lipsol can be extremely variable and, in general, it is safest to use fresh slides, in which case a few seconds exposure may be enough for G-bands to be obtained. Older slides require longer exposure times. Although other methods can certainly be used, care should be taken that a large part of the DNA is not lost in the subsequent steps.

(ii) *Post-banding.*

G-bands. The most widely used method is that developed by Harper (details in *Table 1*) (4). Direct G-banding is observed but the banding is enhanced by a series of de-staining steps followed by re-staining.

Q-bands. (Table 1). Quinacrine mustard has been used successfully to obtain bands (10). Usually, two photographs are necessary as the silver grains do not show up optimally under fluorescent light.

R-bands. (Table 1). Very clear R-bands produced by acridine orange after heat treatment can be produced after *in situ* hybridisation (1).

Other methods. Various other techniques work well in experienced hands (11 – 14) and some investigators have avoided banding altogether by using a marker chromosome (e.g., 14p+, or fragile X) (3,15,16) and demonstrating that the total number of grains falling on this particular chromosome is significantly higher than random background.

2.6 Denaturation, Hybridisation and Wash Conditions

Prior to denaturation, the slides are treated with a dilute solution of RNase to remove any RNA molecules which might potentially compete for hybridisation. This, and the subsequent procedures, are most conveniently carried out by floating the slides on a metal tray in a water bath at the desired temperature.

Chromosomal DNA is usually denatured by heating in a solution containing formamide (this lowers the melting temperature of the DNA strands). A number of slightly different protocols have been used (summarised in *Table 2*).

Hybridisation is carried out in a formamide buffer so that the temperature of hybridisation can be lowered. The formamide should be of the highest quality and should be deionised and then stored at $-20°C$. The intensity of the signal relies on the two strands of DNA forming networks of single strand molecules (so-called Christmas trees). This phenomenon is greatly enhanced by the presence of 10% dextran sulphate in the hybridisation mix, which reduces the effective volume of solution available to the macromolecules and therefore increases their effective concentration. In solution hybridisations it has been shown that polyethylene glycol (PEG 6000) provides a cheaper alternative. Various additions to the buffer have been suggested to block random background hybridisation but it is doubtful whether any of these make any very substantial difference. However, addition of a non-competitive carrier DNA, such as sonicated salmon sperm DNA, to block non-specific hybridisation, is advised. Very little can be done to increase the amount of hybridisation and the signal obtained, so it is important to wash as thoroughly as possible to lower any background. A rather stringent wash in a typical protocol is given in *Table 2*.

Table 2. Treatment of Slides for *In Situ* hybridisation.

A. *RNase treatment*

1. Incubate the slides for 1 h with 200 μl RNase (100 μg/ml in 2 × SSCᵃ). The stock RNase solution (10 mg/ml in 2 × SSC) should be boiled for 3 min and then cooled rapidly to remove any contaminating DNases.
2. Wash the slides thoroughly in 2 × SSC.
3. Dehydrate through a series of alcohols, 50%, 75%, 95%, 100%.

B. *Denaturation*

1. Denature the slides in 70% formamideᵇ/2 × SSC at 70°C for 2 min.
2. Wash quickly in 2 × SSC.
3. Dehydrate through a series of alcohols, 50%, 75%, 95%, 100%.

C. *Hybridisation*

1. Dissolve the probe in 20 μl of hybridisation buffer per slide. Hybridisation buffer is 50% formamide, 10% dextran sulphate, 0.025 M NaCl, 0.04 M sodium phosphate, 0.03 M sodium citrate (pH 6.0).
2. Boil the dissolved probe for 2 min.
3. Apply 20 μl of sample to the slide, cover with a coverslip and seal with rubber solution.

D. *Washing*

1. Gently remove the coverslip.
2. Rinse the slides briefly in 2 × SSC, at room temperature.
3. Wash the slides in 2 × SSC at 55°C for 1 h.
4. Wash the slides in three changes of 50% formamide/2 × SSC at 39°C for 15 min each.
5. Wash the slides extensively (overnight) in 2 × SSC, in the cold.

ᵃSSC is 0.15 M NaCl, 0.015 M sodium citrate.
ᵇFormamide is probably carcinogenic and teratogenic and should be handled with gloves.

2.7 Probe Preparation

2.7.1 *Source of Nucleic Acid*

DNA probes from recombinant plasmids or phage should be of the highest purity available, and should preferably have been purified through a CsCl gradient. There is no need to separate the cloned sequence from the vector by restriction enzyme digestion and, indeed, the extra DNA supplied by the vector may contribute favourably to the DNA networks described in Section 2.6. It is strongly recommended that the probe is well characterised with respect to copy number in the genome and absence of repetitive sequences. If any repetitive sequences are present, they must be removed by sub-cloning.

2.7.2 *Labelling of Nucleic Acid*

Nick-translation is the commonly accepted method of labelling, and all the components for this reaction are readily available from biochemical companies specialising in molecular biology products, either in kit form or individually. Care should be taken to optimise the reaction conditions to give the maximum possible specific activity. Using [³H]deoxynucleotides, specific activities of more than 10^8 d.p.m./μg can be obtained.

Synthesis of a second DNA strand from single-stranded (i.e., boiled) probes using random priming with hexaoligonucleotides is also suitable and gives a higher specific activity product (17).

2.7.3 *Hybridisation Conditions*

It is not clear what the ideal probe concentration for *in situ* hybridisation is, and it probably varies between different probes. Therefore, it is best to use a range of concentrations from 1 ng to 20 ng/slide.

2.8 Analysis of Data

The exact sequence of photography will vary depending on the banding method used (see *Table 1*). Whichever method is favoured the final result will be a set of photographs of chromosomes, with grains scattered about them, for karyotyping. The cells should be karyotyped (at least 20 should be looked at), and then grain positions plotted on an ideogram. This should show a clear concentration of grains at one position of the karyotype which makes statistical analysis unnecessary. Even in the most favourable case, the majority of silver grains observed will not fall over the specific region. The maximum likely to be found at any one position is only about 20%; however, as the other grains will be distributed randomly over the other 99% of the genome, there should be no secondary peak of any significant size. Some workers apply a cut-off in terms of the numbers of grains/cell, but we prefer not to, in order to minimise the degree of bias introduced.

3. MANIPULATION OF SOMATIC CELL HYBRIDS FOR THE ANALYSIS OF THE HUMAN GENOME

3.1 Introduction

The genetic information of cells from very different origins can be combined in a single nucleus by somatic cell fusion. Such hybrid cells have been used extensively in, for example, standard genetic complementation tests and the production of monoclonal antibodies, though it is perhaps in the field of human gene mapping that they have found their broadest application. The concept has recently been extended to the mapping of anonymous DNA sequences isolated through recombinant DNA technology.

When rodent and human cells are fused the resultant hybrids, as they proliferate, selectively eliminate human chromosomes. This process is random and may be so variable that the majority of human chromosomes are lost from some hybrids whilst in others a large number are retained. After some time in culture these hybrids become more stable and further losses are gradual. There is, however, no way of predicting what will happen. Thus, throughout the production and propagation of these hybrids, constant chromosome checks and sub-cloning must be undertaken. In this section the production and subcloning of hybrids is discussed. In addition, the analysis of the chromosomes from rodent/human hybrids is illustrated.

3.2 Selection Systems

Only a small proportion of the cells involved in the fusion yield hybrids. The remainder must be eliminated, otherwise they will overgrow the cultures. A selection system is therefore necessary. The HAT selection system is by far the most commonly used.

Figure 1. The HAT selection system.

3.2.1 *HAT Selection*

HAT stands for the component parts of the system; hypoxanthine, aminopterin, thymidine.

DNA precursors are normally synthesised from glutamine and dUMP. If this pathway is blocked the cells cannot synthesise DNA and will eventually die. Aminopterin blocks this pathway. However, if these blocked cells are supplied with hypoxanthine and thymidine they can generate DNA precursors *via* a salvage pathway using the enzymes hypoxanthine phosphoribosyltransferase (HPRT) and thymidine kinase (TK), respectively (*Figure 1*). Thus normal cells grow quite happily in HAT.

By contrast, cells deficient in HPRT or TK cannot produce DNA precursors in the presence of HAT and are killed. Thus, fusion between TK$^-$ and HPRT$^-$ cells results in complementation within the hybrids which are the only cells to survive.

3.2.2 *Selection and Maintenance of HAT-sensitive cells*

Normal cells will incorporate 6-thioguanine, which is a purine killer analogue, through the HPRT pathway, resulting in their death. Cells deficient in HPRT are therefore resistant to thioguanine (TG) but sensitive to HAT. Cells deficient in HPRT, as a result of induced or spontaneous mutation, can be selected in a single step by growing $5 \times 10^6 - 10^7$ cells on 20 μg/ml TG. After 2 – 3 weeks resistant colonies should appear and must be propagated continuously in TG.

In the same way, the pyrimidine killer analogue BrdU is processed by TK. Thus, TK$^-$ cells are resistant to BrdU but sensitive to HAT. TK$^-$ cells can be isolated by a single step selection in 50 μg/ml BrdU. Again, resistant cells are best propagated on BrdU to avoid the possibility of reversion.

All mutant cell lines should be tested periodically, or at least before each fusion experiment, to confirm their HAT sensitivity.

3.2.3 *Ouabain Selection*

Selection of TK$^-$ and HPRT$^-$ cells takes up to 6 weeks, and large numbers of starting cells are required to ensure the isolation of the mutants. When using primary human cells such as skin fibroblasts or lymphocytes it is not necessary to generate mutant cell lines. Instead, ouabain can be used for selectively killing the unfused human cells. Ouabain poisons the sodium-potassium pump in the cell membrane preventing normal intracellular regulation of these electrolytes. Human cells are 100 – 1000 times more sensitive to ouabain than are rodent cells. Since ouabain resistance is a dominant trait, rodent/human cell hybrids have the same resistance as the parental rodent cells. Thus, during the generation of rodent/human cell hybrids the human parental cells can be eliminated in the absence of TK or HPRT deficiency mutants. The rodent cell line, however, should be HAT sensitive.

Following standard procedures for fusion the selective medium should contain both HAT and ouabain at the appropriate concentration. Ouabain death is characterised by large swollen cells whereas HAT death produces small dense cells. Since the ouabain kills resting and cycling cells it is not necessary to keep the cells in it for longer than 2 weeks. It is necessary, however, to determine the relative sensitivities of the two parental cell lines to ouabain before carrying out the fusion as outlined below. The starting solution for the test series should be 10^{-2} M (*Table 3*) from which appropriate dilutions can be made. In all cases it is sufficient to use distilled water for the dilutions.

Set up a series of 6 × 50 mm plates containing 5 × 10^5 cells from each parental cell line. Add ouabain to a final concentration of between 10^{-3} and 10^{-8} in 10-fold dilution steps. It will be noted that the lower concentrations do not significantly affect growth of either rodent or human cells. At higher concentrations the human cells begin to die, whereas the rodent cells will probably survive at the highest concentration. The concentration to use in the fusion experiment is the one just higher than that required to cause complete killing. In our laboratory we find that 10^{-6} M ouabain kills human cells, whereas 10^{-3} M does not kill rodent cells. We tend, therefore, to use a $5 × 10^{-5}$ M final concentration in the fusion experiments. It is important to be aware that ouabain resistance can occur in human cells, and if presumptive hybrids appear which are morphologically similar to the parental cells they should be subjected to careful karyotypic analysis. Whenever new stocks of ouabain are made it is always advisable to check the effectiveness of the new stock against human cells. Some fibroblast cultures, for instance, are more sensitive to ouabain than others.

3.2.4 *Preparation of HAT*

The details of HAT preparation are given in *Table 3*. Bulk solutions of HAT (500 × and 100 ×) can be stored indefinitely at $-70°C$. In our laboratory HAT is made up in serum-free medium (SFM). As with all our solutions, stocks are 100 × concentrated and whilst in use kept in 10 ml aliquots at 4°C. Thus, 100 μl is added to each 10 ml culture.

We prefer to add the HAT to each dish separately but bulk media preparations can be made containing HAT at the correct concentration and stored either frozen or at 4°C. By adding the selective agents separately the number of stock media bottles is

Table 3. Preparation of Stock Solutions for Selective Media.

	Mol. wt.	*Final con-centration*	*Stock preparation*
Hypoxanthine	136.1	13.6 μg/ml	136.1 mg/100 ml
Aminopterin	440.4	0.17 μg/ml	1.76 mg/100 ml
Thymidine	242.2	3.87 μg/ml	38.7 mg/100 ml
6-Thioguanine	167.2	20 μg/ml	200 mg/100 ml
5-BrdU	307.4	30 μg/ml	300 mg/100 ml
Ouabain[a]	728.6	$5 × 10^{-5}$ M	36.5 mg/100 ml

[a]See text for optimum working concentrations.

kept to a minimum and space is saved.

Thymidine is readily soluble in water but at high concentrations hypoxanthine and aminopterin are not. The addition of 5 M NaOH dropwise during the preparation of the concentrated HAT solutions increases solubility. All of the ingredients can be mixed together and made up to half volume in SFM. At this point just enough NaOH should be added to remove the precipitate. If the HAT is made up in SFM it will be bright pink. When the hypoxanthine and aminopterin are fully dissolved, the solution can be made up to the final volume. Some workers prefer to neutralise the solution with 1 M HCl before making it up to the final volume but we find this unnecessary. On storage, the hypoxanthine and aminopterin may come out of solution but can be returned by addition of more NaOH. All HAT solutions should be filter-sterilised before use. Sometimes the hypoxanthine, thymidine and aminopterin are made separately and added together just before use.

Despite the high concentrations of aminopterin, HAT medium will support fungal growth, and careful monitoring of stock solutions which are used repeatedly over long periods should be undertaken. If in doubt, throw it out. Failure to take this simple precaution can ruin months of hard work.

3.3 Cell Fusion

Depending on the type of cells being used in the experiment, two different fusion protocols are available involving monolayers or suspensions. The original method of fusion, using inactivated Sendai virus, is complex (18,19) and has been superceded by a quick and easy method using polyethylene glycol (PEG).

3.3.1 *Monolayer Fusion*

The procedure outlined in *Table 4* can be used for any cells which show anchorage-dependence. The cells in the dish should form a confluent layer to maximise the number of cells fused. If one of the parental cell lines recovers slowly following trypsin treatment, then it should be seeded first and grown to subconfluence before addition of the other parental cells. Pre-fusion washing removes excess serum which would otherwise precipitate the PEG. For the same reason stock PEG solutions are made up in SFM. The amount of PEG added should be just enough to cover the surface of the cells, 0.5 ml per 5 cm dish is enough. Once added, the PEG should be washed over the surface of the cells by gently rocking the Petri dish. PEG is a detergent and therefore toxic to

Table 4. Protocol for Monolayer Fusion.

1.	Seed 10^6 cells from each parent into a 5 cm Petri dish.
2.	Allow the cells to settle overnight.
3.	Next morning, wash the monolayer three times with PBSA[a].
4.	Add 0.5 ml of 50:50 PEG:SFM and leave for 30−90 sec.
5.	Dilute the PEG solution with excess SFM and remove.
6.	Wash the monolayer gently with three changes of SFM.
7.	Add medium and serum and leave to recover for 4−6 h.
8.	Split each 5 cm dish into 6 × 9 cm dishes.
9.	Add selective agents, e.g., HAT, ouabain.
10.	Change medium twice weekly.

[a]PBSA is phosphate-buffered saline without calcium or magnesium ions.

cells. It is possible that different cells show different sensitivities to PEG, necessitating different treatment times. In our laboratory each fusion experiment is set up in triplicate and treated for 30 sec, 60 sec and 90 sec, respectively. When cell numbers are limited we generally find that 60 sec is sufficient to generate adequate numbers of hybrids. Post-fusion washing is also important, not only to remove the PEG but to prevent precipitation of PEG when serum is added. During the recovery period the cell membranes reform, and it should be possible to see many multinucleate cells in the culture dish. Splitting the cells to low density means that when the hybrids arise they form discrete colonies which makes isolation easier. Since HAT only kills dividing cells, regular changing of the medium encourages growth and hence killing of the parental cells.

3.3.2 *Suspension Fusion*

When one or both of the parent cell lines are anchorage-independent, e.g., lymphocytes, lymphoblastoid cell lines or spleen cells, the fusion must be carried out in suspension as outlined in *Table 5*. Again, pre- and post-fusion washes are important to remove serum. In our laboratory, the fusion is carried out in 25 ml plastic Universal tubes with a conical bottom. Agitation is achieved by fairly hard flicking of the bottom of the tube during exposure to PEG which keeps the cells in contact with the fusing agent for the whole period. Use of large tubes means that dilution is more efficient after fusion. If cell numbers are not limiting then a range of PEG times is suggested.

A slight modification to this procedure which appears to improve the fusion efficiency is to add PHA (12.5 μg/ml) to the cells for $20-30$ min followed by a brief wash in SFM prior to fusion. During this time the cells bind together ensuring contact during the PEG treatment.

If one of the parental cells is anchorage-dependent and the other is not, the hybrids often assume the dependent phenotype. However, it is wise to retain the unattached cells from the early stages of these experiments in case the hybrids assume the anchorage-independent phenotype. It will soon be obvious which is the case. If both parents are anchorage-independent then the fusion mixture should be added to 500 ml of medium following washing, and then aliquotted into multiwell dishes in 1 ml volumes.

3.3.3 *Preparation of PEG*

Stock solutions of PEG (molecular weight 1000) should be sterilised by autoclaving,

Table 5. Protocol for Suspension Fusion.

1.	Harvest both parental cell lines separately.
2.	Mix 5×10^6 cells from each parent and co-sediment.
3.	Wash the mixture of cells three times in excess PBSA.
4.	After the final wash remove all liquid from cells.
5.	Gently resuspend the pellet by flicking tube.
6.	Add 0.5 ml of 50:50 PEG:SFM for $30-90$ sec.
7.	Agitate constantly during treatment.
8.	Dilute the PEG with excess SFM.
9.	Recover the cells by centrifugation.
10.	Wash the cells three times in PBSA or SFM
11.	Distribute at low density depending on cell types used (see text).

then made up to working strength with sterile solutions. PEG should always be stored in the dark, either in a dark bottle or wrapped in silver foil, preferably in small aliquots. The PEG may be repeatedly melted and set either in an autoclave or microwave oven. Some workers make up the 50% PEG solution and freeze it between experiments. We prefer to make the PEG fresh each time. The working solution is made by adding SFM and PEG in equal volumes. If the PEG goes cloudy there is probably serum in the medium. The PEG will remain liquid at 65°C and should be diluted with SFM at 37°C. Mix the PEG thoroughly. Once diluted the 50% solution will not set at room temperature. The SFM also acts as a pH indicator, and when the PEG is added it should remain in the red end of the spectrum. Some workers prefer the PEG to be more acid (yellow) but we strongly recommended that it be used at neutral pH.

3.3.4 *Removing Cells from HAT*

Since the HPRT and TK genes are on chromosomes X and 17, respectively, these chromosomes will be retained in an HPRT$^-$ and TK$^-$ background. If these chromosomes are not of interest in the experiment the hybrid cells can be removed from HAT. This process should follow a gentle weaning through several weeks of hypoxanthine and thymidine supplement. During this time the cells will be able to re-establish the conventional pathway for synthesis of DNA precursors. If, when the cells are removed from hypoxanthine and thymidine, they look 'sick' or fail to thrive they should be returned to HAT. It is generally not a good idea to remove cells from HAT too early since parental cells merely suppressed by HAT may re-emerge and go on to dominate the culture. This use of hypoxanthine and thymidine is the reason why some workers prefer to make up hypoxanthine, thymidine and aminopterin separately.

3.4 **Hybrid Isolation Following Fusion**

Rodent/human cell hybrids are highly unstable, at least in the first instance. During the time it takes to grow up sufficient quantities to permit chromosome analysis many human chromosomes will already have been lost. Since this loss is random, the progeny from each individual hybrid cell are likely to be chromosomally quite different from each other and to have different growth rates. To prevent the better adapted hybrids from dominating the cultures, strategies should be designed to allow isolation of individual hybrid clones separately. The same techniques can be used for recovering particular clones from a stock hybrid culture with chromosome instability.

The first important step is to seed the fused cells out at low density allowing the hybrids to establish their own discrete colonies. At this stage Petri dishes rather than flasks are preferred. The appearance of the hybrids depends on the parental cells, and in our experience hybrids have characteristics of both parents (20). Hybrid colonies looking exactly like either parent should be viewed with caution. Depending on the growth rate of the parental cells, hybrids should be visible in the dishes 2 − 3 weeks after fusion.

3.4.1 *Ring Cloning*

Hybrids of anchorage-dependent parents can be isolated by ring cloning. In this procedure individual hybrid colonies are ringed on the underside of the dish with a marker pen using low power on an inverted microscope, by inserting the pen tip between the underside of the dish and the objective. Cloning rings are cylinders of varying diameter made out of either glass or metal. It is important that the edges of the cylinders are ground smooth. It is convenient to have cloning rings with different diameters (0.5 – 1.5 cm) to cover different sized colonies.

(i)   Prior to cloning, wash the cells with PBSA and rinse.

(ii) Coat one of the open surfaces of the cloning ring with silicone grease. This is achieved by pressing it into a 0.5 cm thick layer of grease in a Petri dish (sterilised by autoclaving) using a pair of sterile forceps.

(iii) Wipe off excess grease on the side of the dish before quickly positioning the ring over the marked areas. If there are a lot of rings in the dish it is particularly important to work quickly to prevent the cells from drying. It is also important, therefore, to deal with one plate at a time. Rings should be tapped firmly in place with the forceps.

(iv) When all colonies have been covered, quickly add one drop of trypsin/versene to each well and leave until the cells begin to detach.

(v)   Inactivate the trypsin with two drops of neat serum.

(vi) Remove each colony independently using a new pipette to avoid cross-contamination. Fill a 1 ml pipette up to the 0.8 ml line and remove all of the fluid in the well into it. Flush up and down gently, avoiding spilling the medium over the edge of the ring.

(vii) Transfer each hybrid into a separate slot in a multiwell dish. If the colonies are large then 24-well dishes are used, while 96-well dishes are preferred for smaller colonies.

(viii) Make up each well to the recommended volume and add the required selective reagents.

(ix) Repeat until all clones are removed.

(x)   During hybrid isolation do not concentrate on the larger, faster growing colonies since the smaller ones are likely to have different chromosome complements, and may have the chromosome(s) of interest. It is generally, but not always, true that the faster growing cells usually have fewer human chromosomes.

The above procedure does not require centrifugation during the first isolation since the trypsin is adequately inactivated by the serum. When the hybrids fill the first dish they should be transferred to the next one up in the series: 96-well, 24-well, 5 cm, 9 cm. Avoid oversplitting the cells in the early stages. Sometimes individual colonies do merge together and cannot be isolated by ring cloning. An alternative to ring cloning requires selection of individual cells under the microscope. In this case the hybrids will be more likely to show a more constant chromosome complement within the population. This technique can also be applied to the selection of particular subclones from a well established hybrid cell line. For example, if a given hybrid contains a single human chromosome, but in only a proportion of the cells, then pure populations of cells with that chromosome can be recovered.

3.4.2 *Individual Cell Cloning*

Perhaps the only sure way to get single cell clones is to pick them individually under a microscope.

(i)   Wash down an inverted microscope with 70% alcohol and place it in a laminar flow hood.

(i)   Seed a 9 cm Petri dish with about 50 cells and place it on the microscope stage.

(iii)   Swirl the fluid in the dish so that there is only a single cell in view under a low power objective.

(iv)   Remove the single cell with an automatic pipette (P20). To prevent large volumes of fluid and hence more than one cell being drawn up, set the pipette to 5 μl and depress the plunger before breaking the surface of the liquid. Position the opening of the tip by the side of the cell to be picked and release the plunger. It should be possible to see the cell disappear into the tip.

(v)   Dispense this cell into a well of a 96-well plate already containing the appropriate medium.

(vi)   Repeat this operation as necessary.

The number of cells in the cloning dish is critical. If there are too many then more than one cell will be removed each time. Also, it is recommended that a non-tissue culture dish is used to prevent the cells attaching during the cloning. With practice this is a quick method of cloning, but it is likely that the medium will turn alkaline and harm the cells. In our laboratory duplicate pairs of dishes are used and alternated between the cabinet and the incubator so that neither stays too alkaline for too long. For best results only select cells which obviously have an intact cell membrane.

Some cells may not clone easily as single cells. We have found that good results are also obtained by cloning cells from 'mitotic blow offs'. Taking a culture with actively growing cells, wash violently over the monolayer with a 10 ml pipette which preferentially removes the mitotic cells. If these are then cloned, then each well effectively starts off with two cells.

Because of the design of the flow cabinet it is sometimes not possible to put the microscope into it. An alternative way of cloning is to seed $30-50$ cells in a 9 cm dish and allow them to attach taking care to keep the dish on a level surface. Each cell should form a discrete colony where it is attached. These colonies can then be removed by ring cloning as outlined earlier.

3.4.3 *Cloning by Dilution*

Single cell cloning can also be achieved by dilution. For anchorage- independent cells this is the only alternative to picking single cells. Count the number of cells available on a haemocytometer or Coulter counter. Dilute the cells until there is 1 cell/ml. For convenience, 100 cells are placed in 100 ml of medium and then dispensed in 1 ml aliquots into a 96-well dish. Clearly, the starting suspension should be of single cells, and after dispensing the cells, the wells should be examined individually and any with more than one cell identified. This method can also be used for anchorage-dependent cells. Following cloning, any well with two foci of growing cells is best presumed to be the result of seeding two cells into the well.

3.5 **Gene Amplification**

Selection of hybrids using the HAT selection system is usually reliable but there is always the risk of reversion in the parental cells. Parental cells becoming TK$^+$ or HPRT$^+$ by 'back mutation' will usually overgrow the hybrid cells. Another mechanism by which parental cells can escape HAT selection is by amplification of the dihydrofolate reductase (DHFR) gene, the target enzyme for aminopterin. Raising the HAT concentration only serves to increase the resistance (21). Chromosome analysis of these resistant cells usually demonstrates either double minutes (DMs) or homogeneously staining regions (HSR) which are the sites of gene amplification (22). It is usually advisable to check the parental stocks from time to time to assess their HAT sensitivity. If, for whatever reason, more than one or two revertants arise, it is advisable to discard that particular stock and recover another from liquid nitrogen.

3.6 Selection of Parental Cells

At the outset, some careful thought about the eventual uses of the hybrids can influence the choice of the parental cells. If the human parental cells are skin fibroblasts or lymphocytes then it is the selection of the rodent cells that is important. The following points should be considered.

(i)   Growth rate. As a rule, the faster the cells grow, the faster the hybrids will grow. This will allow large numbers of cells to be available for analysis quickly. The faster growing cells also show a greater cloning efficiency which is useful if sub-cloning becomes important.

(ii)   Chromosome number. For certain experiments such as the production of monoclonal antibodies, chromosome number is not particularly important. When detailed chromosome analysis must be done, however, the fewer chromosomes from the malignant parent the better. This is particularly important when using DNA probes for mapping since the 'signal' produced by the human component of the hybrid is reduced as the amount of non-human DNA increases.

(iii)   The karyotypes of the parental rodent cell should be as near normal as possible. For example, in the mouse all of the normal chromosomes are telocentric. Thus in mouse/human hybrids only the D-group of human chromosomes are potentially confused. However, if the mouse parent has a grossly rearranged karyotype such that large numbers of metacentric mouse chromosomes have been generated, it becomes more important to be able to identify all of the mouse chromosomes unequivocally. Anyone not familiar with hamster chromosomes might find hamster/human hybrids difficult to analyse since the hamster has a chromosome size and centromere distribution somewhat similar to the human karyotype.

3.7 Mycoplasma Contamination

Mycoplasma are intracellular parasites which attach on the inside of the mammalian cell membrane. Cells contaminated with mycoplasma will not fuse. Any experiment in which all of the cells die within the first 7−10 days should be suspected of being contaminated. Usually it takes 2−3 weeks before HAT kills most of the cells and often healthy looking cells remain throughout the experiment, albeit in a dormant state. The best method for detecting mycoplasma is to culture it on selective medium, which will identify even low level contamination. Alternatively, staining with Hoechst 33258 stain will highlight the A+T-rich DNA of the mycoplasma which will appear as bright extranuclear spots in the cytoplasm. Heavy contamination appears like strings of fluorescing material, again in the cytoplasm.

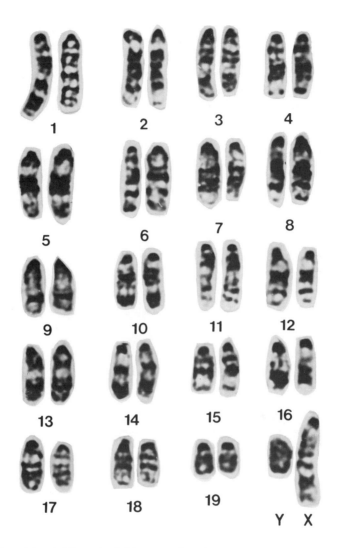

Figure 2. Normal trypsin-Giemsa banded karyotype from the mouse.

3.8 Chromosome Analysis of Hybrids

The majority of rodent/human hybrids are constructed with either mouse or hamster cells. In some cases a detailed knowledge of human chromosomes alone might be sufficient to analyse them if there are no rodent/human translocations, and the parental rodent cells themselves do not have any rearrangements resembling human chromosomes. However, accurate analysis of these hybrids requires a detailed knowledge of the chromosomes from both parents.

3.8.1 *Mouse Chromosomes*

The mouse has 40 chromosomes, and in normal cells they are all telocentric (*Figure*

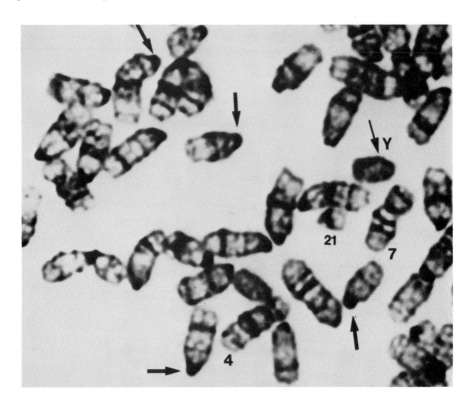

Figure 3. Partial karyotype from a mouse/human somatic cell hybrid. The mouse chromosomes have darkly stained centromeres (indicated by the arrows) with the exception of the Y chromosome. Human chromosomes are identified by their numbers. Note the darkly staining centromeric region of human chromosome 21 (see text).

2). With the exception of the Y chromosome (*Figure 3*), the centromeric regions of the mouse chromosomes are highly A+T rich. Even with trypsin-Giemsa banding the mouse centromeres stain intensely (*Figure 3*). This feature alone is often sufficient to identify the mouse chromosomes. The centromeres from some human metacentric chromosomes also stain darkly but the telocentric ones do not. The high A+T content of mouse centromeres means that they also stain intensely with the fluorochrome Hoechst 33258. Whilst this technique will identify the number of human chromosomes present it will not identify the individual chromosomes since the banding produced is not very distinct.

Similarly, G-11 banding will indicate the number of human chromosomes present and in particular will also identify any mouse/human translocations. The G-11 banding technique was designed by Bobrow and Cross (23) to distinguish between human and rodent chromosomes. The staining is carried out in Giemsa at pH 11. The protocol for G-11 staining is given in *Table 6*. This procedure highlights specific regions of constitutive heterochromatin in the human karyotype. The human chromosomes stain bluish with red differentiation of paracentromeric regions of chromosomes 1, 4, 5, 7, 9, 10, 13, 14, 15, 20, 21 and 22. The rodent chromosomes stain dark magenta with

Table 6. Protocol for G-11 Staining.

1.	Age chromosomes 7 days before staining.
2.	Prepare a stock solution of 0.7 M NaOH
3.	Dilute the NaOH stock 1:50 in distilled water.
4.	Dilute Giemsa 1:20 in dilute NaOH stock solution.
5.	Treat the slides in 2 × SSC for 5 min at 56°C.
6.	Stain for 20 min in the diluted Giemsa.
7.	Rinse briefly in PBSA (pH 6.8) and dry.

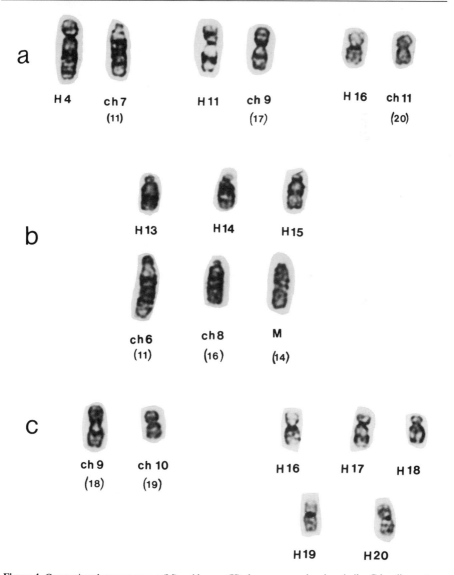

Figure 4. Comparison between mouse (M) and human (H) chromosomes showing similar G-banding patterns.

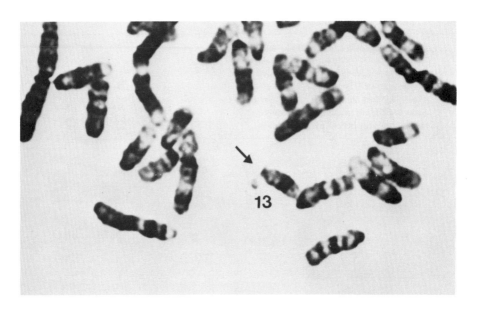

Figure 5. Mouse/human hybrid showing human chromosome 13; note the absence of a darkly staining centromere compared with the mouse chromosomes.

pale blue centromeres. As with most chromosome banding techniques the results can be variable, although, in general, the human chromosomes stain lighter than the rodent ones. Standard chromosome preparations can be used although it is advised that they are allowed to age for 7 days before use. It was shown by Bobrow and Cross that, following G-11 staining, the human chromosomes could then be identified by quinacrine banding. The chromosome spreads should, however, be dilute on the slides otherwise the stain will concentrate around the edges of the preparation rather than throughout it (D.Sheer, personal communication). It also appears that the type of stain used is important. Bobrow and Cross suggested the Harleco Company for the source of Giemsa; BDH stains have also been used successfully, whereas Gurr stains are generally not so good. The large variation in stain components between batches from different manufacturers means that it is likely that some batches will work better than others. The best advice is to find one that works and try to stick to it.

It is perhaps advisable for those unfamiliar with rodent chromosomes first to determine the number of human chromosomes expected in the hybrids, and then to identify the same number by G-banding.

A detailed consideration of the banding patterns of the mouse chromosomes has been published previously (24). Being telocentric the only potential confusion is with the D- and G-group human chromosomes and the Y chromosome. In both species the Y chromosome does not have an obvious centromere and stains uniformly grey. If, therefore, the Y chromosome is the subject of the experiment, female rodent cells should be used in the fusion.

Human chromosome 13 is similar to mouse chromosome 6 (*Figure 4a*) and the human chromosome 14 resembles mouse 10 (*Figure 4b*). In both cases the G-negative

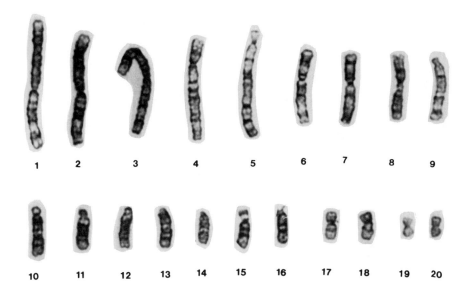

Figure 6. Trypsin-Giemsa banded chromosomes from Chinese hamster cell line WG1E. The chromosomes are arranged in order of decreasing size and numbered consecutively. These numbers do not relate to the standard nomenclature for hamster chromosomes.

characteristics of the human centromeres together with the characteristic appearance of the ribosomal genes on the short arm of human 13 and 14 should be sufficient to distinguish them from the mouse chromosomes (see *Figure 5*). In the G-group of human chromosomes, human 22 resembles mouse 19 (*Figure 4b*) although in well banded preparations the mouse 19 has two centrally located bands whereas the human 22 has only one. The mouse 19 is also generally bigger than the human 22. Since the centromere of human chromosome 22 usually stains darkly with Giemsa this is not a feature for distinguishing it from mouse chromosome 19. The prominent satellite on the short arm of human 22, however, is useful in this respect (see *Figure 4b*).

Often rodent-human hybrids are created to isolate particular abnormal human chromosomes for gene mapping. In these cases certain regions of the human chromosomes resemble mouse chromosomes. The long arm of human chromosome 1 can be confused with mouse chromosome 6 (*Figure 4a*). The long arms of human chromosomes 11 and 12 are also similar to mouse chromosomes 7 and 5, respectively (see *Figure 2*). Similarly, the long arm of human chromosome 16 is similar to mouse 18. It is not possible to compare all of the regions of all the chromosomes in the two species but the most obvious similarities are given in *Figure 4*. The best advice is to select a mouse parent with only telocentric chromosomes which makes the analysis a lot easier. It is worth remembering that if the human abnormality is very subtle, such as a small deletion, in the absence of the normal homologue for comparison purposes it may be difficult to determine which of the two homologues is present.

3.8.2 Hamster Chromosomes

Normal Chinese hamster cells have 22 chromosomes of which, unlike the mouse, a

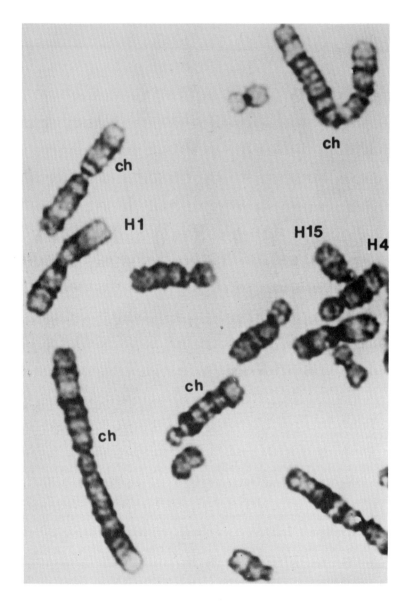

Figure 7. Partial metaphase spread from a human/hamster somatic cell hybrid. Note the relative sizes of the larger hamster chromosomes (ch) compared with human chromosome 1 (H1).

detailed analysis does not exist (25,26). A popular variant of the normal karyotype is shown in *Figure 6* and demonstrates the main features. Firstly the hamster karyotype has a similar range of chromosome sizes to the human karyotype, with large and small metacentrics and telocentrics. Because of the many marker chromosomes in this cell line (WG1E) they are merely numbered in decreasing order of size. The numbers in *Figure 6* have no relationship to those given to the normal karyotype by Kato and Yoshida

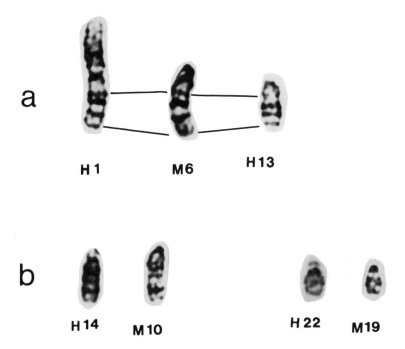

Figure 8. Size comparison between hamster (ch) and human (H) chromosomes from the same metaphase spread. The numbers in brackets below the Chinese hamster chromosomes refer to Figure 6.

(25). Unlike the mouse chromosomes, the centromeric regions in the hamster are not AT rich and so Hoechst 33258 cannot be used. G-11, however, will distinguish between hamster and human chromosomes. The larger hamster chromosomes are far bigger than the largest of human chromosomes and can be excluded from the analysis (*Figure 7*). Others, however, are very similar (*Figure 8*).

In *Figure 8* a comparison of hamster and human chromosomes is presented from the same metaphase spread. All of the D-group chromosomes are much smaller than the telocentric hamster chromosomes (*Figure 8b*) and the smallest hamster chromosome is about the size of human chromosome 16. This observation can be important in the initial analysis of hamster/human hybrids. Although the human 16 and hamster 11 are similar, the darkly staining centromere of human 16 is the most distinguishing feature. The same is also true, although to a lesser extent, for human chromosomes 19 and 20 which also have darkly staining centromeres (*Figure 8c*) although banding patterns on human 20 and hamster 11 are similar. None of the smaller hamster chromosomes have a darkly staining centromere in Giemsa. It can be seen that human chromosome 4 is roughly the same size as the second largest telocentric hamster chromosome, whereas human chromosome 11 is about the same size as the second smallest hamster chromosome (see *Figure 8a*). Hamster chromosome 8 is similar to human chromosome 13 although the most proximal heavy band in the hamster 8 is much bigger than that in human 13.

3.8.3 *Chromosome Techniques*

In the previous chapters of this volume chromosome analysis techniques have been discussed in detail. However, chromosome preparations from hybrid cells may require different strategies from those used to analyse the parental cells. In particular we have found that the hypotonic treatment often requires modification. Hybrid cells contain many more chromosomes than the parental cells and often endoreduplicate. In order to obtain well spread metaphases it might thus be necessary to increase the length of the hypotonic treatment. On the other hand, some cells are more sensitive to hypotonic than others. The particular Chinese hamster cell line discussed earlier requires only 0.1 M KCl for 3 min to produce well spread chromosomes whereas human fibroblasts used in fusions require 10 min in 0.07 M KCl. Hybrids between these two cell types produce well spread preparations after 5 min of 0.1 M KCl. Thus for optimal spreading it pays to determine the exact conditions for each set of hybrids. Excessive hypotonic treatment will cause overspreading of the chromosomes such that it will be difficult to determine whether or not variations in human chromosome complement are due to random loss during spreading or to true variation between cells. Unlike stimulated lymphocytes, hybrid cells grow continuously in culture and so a constant proportion of cells will always be in mitosis. The mitotic cells can usually be seen clearly under a low power inverted microscope. If there are insufficient mitotics — at least 10 per field with a × 20 objective — it is advisable to change the medium on the cells 18−24 h before beginning the preparation. The fresh medium will stimulate the cells to grow, although only in subconfluent cultures. Confluent cultures have to be split. In sparsely populated dishes, although there are a high proportion of mitotics, the number of cells is low and chromosome preparation can be difficult. In particular, pellets of cells are small and the number of slides that can be made at the end is limited.

4. FLOW CYTOMETRIC ANALYSIS OF HUMAN CHROMOSOMES

4.1 **Introduction**

The banding techniques described in Chapter 3 have been extremely successful in the analysis of the human karyotype. This approach, however, is very labour intensive, requiring skilled and patient analysis at the microscope. Any automated system of karyotype analysis would be of great benefit to the cytogeneticist in terms of economy and greater objectivity. In recent years, flow cytometry has shown the greatest potential for providing a high resolution analysis of the human karyotype. This approach is based on the accurate measurement of the fluorescence of a suitably stained suspension of metaphase chromosomes. A flow cytometer is capable of measuring fluorescence with a 1% coefficient of variation and, with such accuracy, it is possible to differentiate between many of the human chromosomes. Chromosomes can be analysed at rates up to 1000/sec and therefore a large representative number of measurements can be accumulated rapidly. A further feature, which has been reported elsewhere (27), is that individual chromosomes can be sorted with high resolution and used for recombinant DNA experiments.

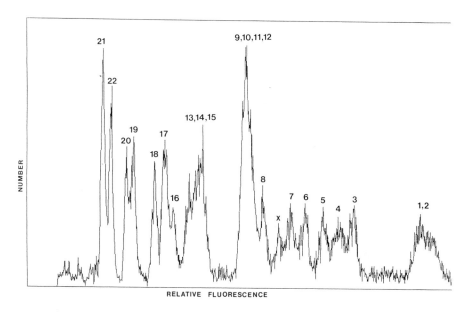

Figure 9. Flow karyotype from human peripheral blood lymphocytes (male). Chromosomes were prepared by the method of Sillar and Young (36) and stained with ethidium bromide. The numbers indicate the peaks produced by each of the human chromosomes.

4.2 **Principles of Flow Cytometry**

In flow cytometry, the objects to be analysed are stained in suspension with an appropriate fluorochrome, and passed singly through a beam of light. The emitted pulses of fluorescence are measured and stored in the form of a histogram displaying the number of pulses *versus* fluorescence intensity. Thus, any population of cellular objects that can be made sufficiently fluorescent can be analysed in this way. Gray *et al.* (28) first analysed Chinese hamster metaphase chromosomes by flow cytometry after staining with ethidium bromide, and obtained a characteristic series of peaks. This was later extended to human chromosomes, and other fluorochromes were introduced (29,30). Subsequently, Young *et al.* (31) extended this approach to allow the improved resolution of peripheral blood lymphocytes to chromosomes on conventional flow cytometers. A typical flow analysis is shown in *Figure 9* with the peaks identified with the corresponding chromosome number. It can be seen that certain chromosomes are quite well resolved, while others, such as 9 − 12, are not.

4.3 **Preparative Techniques**

One of the most important steps in flow analysis is the preparation of a monodisperse suspension of metaphase chromosomes. Although a number of different approaches have been developed over the last few years, they all have certain features in common. A culture of growing cells is usually treated with an agent such as colchicine or vinblastine in order to arrest sufficient cells in metaphase. The period of treatment may depend on the cell cycle time, but extensive culture in the presence of such agents can lead to a high degree of chromosome contraction. If the cells to be examined grow

Table 7. Chromosome Preparation for Flow Cytometry.

pH	Detergent	Lysis	Stabilising agent	Fluorochrome	References
7.5	–	S	Hexylene glycol	EB	Gray *et al.* (28), Stubblefield *et al.* (34)
7.5	Triton X	S	Mg^{2+}	EB+M	Otto and Oldiges (62)
7.2	Digitorin	V	Polyamines	EB	Sillar and Young (36)
–	–	V	PI	PI	Matsson and Rydberg (41)
–	Triton X	S	Psoralen (AMT)	Hoechst 33258	Yu *et al.* (39)
–	Triton X	S	PI	PI	Buys *et al.* (40)
1% Acetic acid	–	U	Hexylene glycol	DAPI	Stoehr *et al.* (43)
7.5	Triton X	S	Mg^{2+}	PI	Bijman (32)
8.0	–	S	Mg^{2+}	PI, Hoechst 33258	Van den Engh *et al.* (47)

S = syringe; V = vortexing; U = ultrasound; EB = ethidium bromide; PI = propidium iodide; M = mithramycin.

as an attached layer, mitotic shake-off can be used to obtain an enriched population of metaphase cells. However, suspension cell lines may also be used if sufficient metaphase cells are present and if care is taken not to lyse the nuclei of the interphase cells which predominate in the population. The cells are then usually subjected to hypotonic swelling and treated with a detergent. The final step is to lyse the mitotic cells by mechanical disruption such as passage through a fine needle or by vigorous vortexing. During the lysis process, it is helpful to monitor the disruption of the mitotic cells on a fluorescence microscope, since insufficient treatment will not lyse an adequate number of cells, or disrupt chromosome clumps, whereas excessive treatment can cause chromosome breakage. The occurrence of breaks in chromosomes, or non-disrupted aggregates of chromosomes, will increase the background level in the flow analysis histogram. An ideal method, therefore, will minimise both the background level and the coefficient of variation of the peaks in the flow karyotype. At the moment, there is no clear argument among investigators as to the best preparative technique for flow analysis, and therefore all the methods used are summarised in *Table 7*. When deciding which method to use, it should be borne in mind that the different methods may be optimal for a particular cell type, as suggested by Bijman (32) or even for a particular type of flow cytometer. The method described in *Table 8* has been used successfully with a large number of cell lines on commercially available cell sorters.

The first flow karyotype analyses (28,33,34) were obtained using the hexylene glycol-based procedure of Wray and Stubblefield (35). This approach has the advantage that the banding structure of chromosomes is sufficiently preserved to allow direct analysis of chromosomes after sorting. Carrano *et al.* (29) investigated the influence of several factors, such as shearing forces and cell concentration, on this protocol. Also, the addition of sodium dodecyl sulphate, ribonuclease and trypsin were found to have a negative effect. An alternative approach for the bulk isolation of chromosomes using polyamines to stabilise the DNA was developed for flow analysis by Sillar and Young (36) (see *Table 8*) from the method of Blumenthal *et al.* (37). A direct comparison of the two methods using the same Chinese hamster ovary (CHO) cell culture indicated that the polyamine method yielded lower background levels and superior coefficients of variation. A disadvantage of this method is that the highly condensed state of the

Table 8. Protocol for Chromosome Preparation for Flow Analysis.

This method of Sillar and Young (36) is based on that described by Blumenthal *et al.* (37) in which the polyamines, spermines and spermidine are used to stabilise the chromosomes and the detergent digitonin is used to assist cell lysis. This method has been used as the preparation technique prior to flow sorting by a number of workers and in our laboratory we have found it to be the most satisfactory.

1. Treat cells with 0.031 μg/ml of colcemid for 5 – 16 h, depending on the rate of cell growth.
2. Harvest mitotic cells by gentle shaking of the culture flasks.
3. Centrifuge at 800 r.p.m. for 10 min.
4. Discard the supernatant and resuspend in 0.075 M KCl at 4°C for 30 min in order to swell the cells.
5. Centrifuge at 300 r.p.m. for 15 min.
6. Discard the supernatant and resuspend the pellet in isolation buffer (15 mM Tris-HCl, 0.2 mM spermine, 0.5 mM spermidine, 2 mM EDTA, 0.5 mM EGTA, 80 mM KCl, 20 mM NaCl, 14 mM 2-mercaptoethanol (pH 7.2).
7. Centrifuge at 1500 r.p.m. for 2 min.
8. Repeat steps 6 and 7.
9. Resuspend the final pellet in approximately 20 times its volume of ice-cold isolation buffer containing 0.1% digitonin. Digitonin is dissolved just prior to use by warming to 37°C and filtering out any insoluble materials.
10. The final volume, typically 2 ml, is agitated vigorously for 1 min on a vortex mixer.
11. The lysis of metaphase cells should be monitored by fluorescence or phase contrast microscopy.

chromosomes prevents direct banding on sorted chromosomes. However, quinacrine studies on sorted chromosomes prepared in this way have recently been described by Fantes *et al.* (38). Yu *et al.* (39) investigated the use of 4'-aminomethyl-4,5',8-trimethylpsoralen to stabilise the chromosomes, but their data did not indicate that this method was superior to the hexylene glycol method.

The use of the intercalating fluorochrome propidium iodide to stabilise chromosome structure has been investigated by Buys *et al.* (40). It was demonstrated that chromosomes isolated in this way maintained their banding structure. A similar protocol but with 0.1% sodium citrate and the addition of 20% ethanol prior to lysis has been reported by Matsson and Rydberg (41). This method is based on that developed by Krishan (42) for staining human lymphoblasts and was claimed to cause less damage to interphase cell nuclei than other methods. Stoehr *et al.* (43) have reported a hexylene glycol-based protocol which involves mild fixation with 1% acetic acid and sonication to disrupt the mitotic cells. Although acidic treatment (50% acetic acid) was shown by Tien Kuo (44) to result in extensive damage to the chromosomal DNA, it was claimed that the relatively mild treatment in this protocol would not prevent DNA analysis on sorted chromosomes. A low pH protocol has been devised by Collard *et al.* (45) for the velocity sedimentation of chromosomes followed by flow analysis. Mitotic cells are lysed in the presence of 2% citric acid which gives superior resolving power on 1 *g* sedimentation. However, since chromosomes sedimented at low pH are not suitable for DNA analysis, Collard *et al.* (46) have subsequently modified the sucrose gradient to neutral pH.

A comparative study on the influence of several ions and detergents has been reported by Bijman (32). The optimal chromosome buffer contained magnesium (3 – 5 mM), sodium (10 mM) and Triton X-100 (0.4%) as judged by flow analysis. Magnesium ions were found to be necessary for the stability of chromosomes whereas sodium and potassium had no effect, with calcium having a negative effect. Recently, Van Den Engh and Trask (47) have described an isolation buffer at pH 8.0 which uses MgSO$_4$ to stabilise the chromosomes. Another feature of this method is the use of RNase to reduce the background fluorescence due to binding of fluorochromes to RNA.

4.3.1 *Choice of DNA-specific Stain*

There are several criteria which a DNA-specific stain must fulfil if it is to be useful in flow cytometry.

(i)   The wavelength of excitation must match the available light source. Most flow instruments use laser illumination and have a restricted set of output wavelengths in the range 458−514 nm and also an ultraviolet output at 360 nm.

(ii)   The relatively small DNA content of most chromosomes requires that the chosen stain has the highest possible quantum efficiency. This is important in reducing statistical errors in measurement of chromosomal fluorescence.

(iii)   The stain should be insensitive to the degree of chromosomal contraction which can be quite variable in chromosomal preparations.

The first flow karyotypes were obtained using ethidium bromide as a DNA-specific stain (28,33,34) and this has remained one of the principle stains used. Jensen *et al.* (48) investigated the suitability of ethidium bromide, Hoechst 33258 and chromomycin A3 for flow cytometric analysis of chromosomes and demonstrated that both Hoechst 33258 and chromomycin A3 yielded flow karyotypes with distinct differences from that obtained by ethidium bromide. It has been shown that Hoechst 33258 has an AT binding preference (49), chromomycin A3 has a GC binding preference (50) and ethidium bromide has no base sequence preference (51). It has therefore been concluded that the differences observed with such stains are due to differences in chromosomal base composition or accessibility. A comparative study of the use of ethidium bromide, Hoechst 33258, chromomycin A3, DAPI and propidium iodide for flow cytometry of metaphase chromosomes has been reported by Langlois *et al.* (52). DAPI binds preferentially to AT-rich DNA and the relative intensities of DAPI-stained chromosomes are similar to Hoechst 33258-stained chromosomes. Similarly, propidium iodide, which is not thought to have a base preference, yields profiles which are similar to those obtained using ethidium bromide. It was also shown by analysis of double stained chromosomes that bound sites interact by energy transfer with little or no binding competition. Conventional fluorescence microscopy was used by Latt *et al.* (53) to study interactions between a number of DNA-binding dyes on metaphase chromosomes. It was shown that if the energy acceptor dye (e.g., actinomycin D or Methyl green) has a binding specificity opposite to the binding or fluorescence specificity of the donor (e.g., Hoechst 33258, quinacrine or chromomycin A) contrast in donor fluorescence can be enhanced, leading to patterns selectively highlighting standard or reverse chromosome bands. The potential of such dye combinations has yet to be fully realised in flow cytometry.

4.3.2 *Technical Innovation*

The quality of flow karyotype analysis is critically dependent on the resolving power of the machine used. Much effort has gone into technical improvement and innovation in order to obtain superior analysis. Conventional machines measure the fluorescence emitted by each chromosome in terms of pulse height or area. However, Cram *et al.* (54) found pulse width to be a good parameter for resolving chromosomes as a function of total emission in the case of the smaller chromosomes and orientation (i.e., arm length) for larger chromosomes. A novel method which is based on chromosome structure, rather than simply DNA content, has been described by Gray *et al.* (55,56). This approach, split scan flow cytometry, is based on the principle that certain chromosomes will orientate in the direction of flow. Therefore, as each chromosome passes a narrow window, a fluorescence signal along the length of the chromosome, with a pronounced drop at the centromere, should be obtained. Such profiles have been obtained from the Indian Muntjac No. 1 chromosome (55,56) and from certain Chinese hamster chromosomes (57) and the centromeric indices thus calculated were found to be in agreement with the value obtained by conventional microscopic measurement. Although this approach has the capability of identifying the larger chromosomes in flow, it has been unable to analyse chromosomes which are too small to orientate in sample stream.

One of the most promising innovations, particularly for the analysis of human chromosomes, has been the introduction of dual beam flow cytometry. Chromosomes are double stained with Hoechst 33258 and chromomycin A3, each of which is excited independently, and the resultant fluorescence values are presented as a two-dimensional histogram using contour plotting. It has been demonstrated that the Hoechst 33258 and the chromomycin A3 fluorescence can each be measured almost independently (55). With this technique, it has been possible to resolve all of the human chromosomes, with the exception of numbers 9 – 12, 14 and 15 (58). A disadvantage of this approach for its general application is that, currently, expensive laser light sources are required.

Recently, high resolution of Chinese hamster chromosomes has been reported (59) using a specially designed flow cytometer to achieve high illumination radiance. It was possible to achieve 10 times greater irradiance than with most other machines. As a result, the fluorescence from metaphase chromosomes was less dependent on laser power, with the doubling of power from 1 to 2 W producing only a 15% increase in signal.

4.3.3 *Flow Karyotype Analysis*

With current techniques and equipment it is possible to measure the fluorescence of human chromosomes with coefficients of variation of 1 – 2%. At such a level of resolution, many of the human chromosomes can be individually resolved and it has therefore become possible to quantify accurately the DNA content in both normal and abnormal karyotypes. These developments have, for the first time, opened up the possibility of using flow cytometry to provide routine karyotype analysis. In order to investigate the feasibility of such an application, we have applied flow cytometry (31) to chromosomes prepared from a series of peripheral blood samples. It was shown that the slight variations in these flow karyotypes could be attributed to polymorphic heterochromatin at the centromeres of certain chromosomes. Flow analysis has shown that certain chromosomes are relatively invariant in DNA content, whilst others, such as the Y chromosome, can vary considerably in DNA content in normal individuals (38). Clearly, before flow cytometry can be fully developed to provide automated analysis of the human karyotype, it will be necessary to take account of these polymorphisms.

The effect of certain chromosomal abnormalities on the flow karyotype can be predicted. For example, a trisomy should be detected as a 50% increase in the area of the peak corresponding to that chromosome. Translocations which result in a net change in DNA content should be detected by the presence of the chromosomes involved in an abnormal position in the profile. Recently, Wirchubsky *et al.* (60) have applied flow analysis to a series of cell lines bearing translocations associated with Burkitt's lymphoma. In particular the 14q+ chromosome in the t(8;14), the 2q+ in the t(2;8) and the 22q− in the t(8;22) could all be separately identified by flow analysis.

Flow cytometry has also proven useful in assessing chromosomal changes occurring in experimental systems. Karyotypic changes have long been observed in spontaneously transformed Chinese hamster embryo fibroblasts and Cram *et al.* (61) have used conventional banding studies and high resolution flow cytometry to analyse in detail such progressive changes. One of the earliest changes, which preceded tumorigenicity in nude mice, was the occurrence of a trisomy of chromosome 5. A steady progression in karyotype instability was observed, corresponding to neoplastic evolution of these cells.

Flow analysis can also be used to monitor random karyotypic changes induced by mutagenic agents. The clastogenic effect of chemical mutagens and X-rays on Chinese hamster chromosomes was monitored by Otto and Oldiges (62) using flow analysis. Such agents cause random damage to chromosomes and therefore result in the broadening of the peaks in the profile. The clastogenic effectiveness of such agents was quantified and the dose-effect relationship was established by the increase of the coefficient of variation of the peak of the largest chromosome type in the flow histogram.

It should be emphasised that there are certain karyotypic changes which would be undetectable by flow cytometry. For example, a reciprocal translocation in which equal amounts of DNA were exchanged would have no effect on the flow karyotype. For this reason flow cytometry is best used in conjunction with conventional banding studies to provide accurate quantitative data on karyotype changes. The main areas for future development in this field will be improvements in resolution and development of staining techniques which overcome variations in fluorescence due to heterochromatic polymorphisms at the centromere.

5. ACKNOWLEDGEMENTS

S.M. has been generously supported while developing these techniques by the Medical Research Council and the Leukaemia Research Fund. J.C. would like to thank Dr J. Pritchard for critical reading of the manuscript. J.C. is supported by a grant from the Imperial Cancer Research Fund.

6. REFERENCES

1. Bartram,C.R., de Klein,A., Hagemeijer,A., van Agthoven,T., Cuerts van Kessel,A., Bootsma,D., Grosveld,G., Ferguson-Smith,M.A. and Cuoffen,J. (1983) *Nature*, **306**, 277.
2. Davis,M., Malcolm,S. and Rabbitts,T.H. (1984) *Nature*, **304**, 286.
3. Boyd,Y., Buckle,V.J., Munro,E.A., Choo,K.H., Migeon,B.R. and Craig,I.W. (1984) *Ann. Hum. Genet.*, **48**, 145.

4. Chandler,M.F. and Yunis,J.J. (1978) *Cytogenet. Cell Genet.*, **22**, 352.
5. Neel,B.G., Jhanwar,S.C., Chaganti,R.S.K. and Hayward,W.S. (1982) *Proc. Natl. Acad. Sci. USA,* **79**, 7842.
6. LeBeau,M.M., Diaz,M.O., Karin,M. and Rowley,J.D. (1985) *Nature,* **313**, 709.
7. Malcolm,S., Barton,P., Murphy,C., Ferguson-Smith,M.A., Bentley,D.L. and Rabbitts,T.H. (1982) *Proc. Natl. Acad. Sci. USA,* **79**, 4957.
8. Manuelidis,L., Langer-Safer,P. and Ward,D.C. (1982) *J. Cell. Biol.*, **95**, 619.
9. Landegent,J.E., Jansen in de Wal,N., Baan,R.A., Hoeijmakens,J.H.J. and Van der Ploeg,M. (1984) *Exp. Cell Res.*, **153**, 61.
10. Kirsch,I.R., Morton,C.C., Nakahara,K. and Leder,P. (1982) *Science (Wash.),* **216**, 301.
11. Bernheim,A., Berger,R. and Szabo,P. (1984) *Chromosoma*, **89**, 163.
12. Cannizzaro,L.A. and Emanuel,B.S. (1984) *Cytogenet. Cell Genet.*, **38**, 308.
13. Zabel,B.U., Naylor,S.L., Sakaguchi,A.Y., Bell,G.I. and Shows,T.B. (1983) *Proc. Natl. Acad. Sci. USA,* **80**, 6932.
14. Popescu,N.C., Arrsbaugh,S.C., Swan,C.D. and Di Paolo,J.A. (1985) *Cytogenet. Cell Genet.*, **39**, 73.
15. Barker,P.E., Rabin,M., Watson,M., Breg,W.R., Ruddle,F.H. and Verma,I.M. (1984) *Proc. Natl. Acad. Sci. USA,* **81**, 5826.
16. Purrello,M., Alhadeff,B., Esposito,D., Szabo,P., Rocchi,M., Truett,M., Masiarz,F. and Siniscalco,M. (1985) *EMBO J.*, **4**, 725.
17. Feinberg,A.P. and Vogelstein,B. (1984) *Anal. Biochem.*, **132**, 6.
18. Harris,H. and Watkins,J.F. (1965) *Nature*, **205**, 640.
19. Marshall,C.J. and Dave,H. (1978) *J. Cell Sci.*, **33**, 171.
20. Cowell,J.K. and Franks,L.M. (1984) *Int. J. Cancer*, **33**, 657.
21. Cowell,J.K. (1982) *Cell Biol. Int. Rep.*, **6**, 393.
22. Cowell,J.K. (1982) *Annu. Rev. Genet.*, **16**, 21.
23. Bobrow,M. and Cross,J. (1974) *Nature*, **251**, 77.
24. Cowell,J.K. (1984) *Chromosoma*, **89**, 294.
25. Kato,H. and Yoshida,T.H. (1972) *Chromosoma*, **36**, 272.
26. Stubblefield,E. (1980) *Cytogenet. Cell Genet.*, **26**, 191.
27. Young,B.D. (1984) *Basic Appl. Histochem.*, **28**, 9.
28. Gray,J.W., Carrano,A.V., Steinmetz,L.L., Van Dilla,M.A., Moore,D.H., II, Mayall,B.H. and Mendelsohn,M.L. (1975) *Proc. Natl. Acad. Sci. USA,* **72**, 1231.
29. Carrano,A.V., Van Dilla,M.A. and Gray,J.W. (1979) in *Flow Cytometry and Sorting,* Melamed,M., Mullaney,P.F. and Mendelsohn,M.L. (eds.), John Wiley and Sons, New York, p...
30. Carrano,A.V., Gray,J.W., Langlois,R.G., Burkhart-Schultz,K.J. and Van Dilla,M.A. (1979) *Proc. Natl. Acad. Sci. USA,* **76**, 1382.
31. Young,B.D., Ferguson-Smith,M.A., Sillar,R. and Boyd,E. (1981) *Proc. Natl. Acad. Sci. USA,* **78**, 7727.
32. Bijman,Th.J. (1983) *Cytometry*, **3**, 354.
33. Gray,J.W., Carrano,A.V., Moore,D.H., II, Steinmetz,L.L., Minkler,J., Mayall,B.H., Mendelsohn,M.L. and Van Dilla,M.A. (1975) *Clin. Chem.*, **21**, 1258.
34. Stubblefield,E., Cram,S. and Deaven,L. (1975) *Exp. Cell Res.*, **94**, 464.
35. Wray,W. and Stubblefield,E. (1970) *Exp. Cell Res.*, **59**, 469.
36. Sillar,R. and Young,B.D. (1981) *J. Histochem. Cytochem.*, **29**, 74.
37. Blumenthal,A.B., Dieden,J.D., Kapp,L.N. and Sedat,J.W. (1979) *J. Cell Biol.*, **81**, 255.
38. Fantes,J.A., Green,D.K. and Cooke,H.J. (1983) *Cytometry*, **4**, 88.
39. Yu,L.C., Aten,J., Gray,J. and Carrano,A.V. (1982) *Nature*, **293**, 154.
40. Buys,C.H.C.M., Koerts,T. and Aten,J.A. (1982) *Hum. Genet.*, **61**, 157.
41. Matsson,P. and Rydberg,B. (1981) *Cytometry*, **1**, 369.
42. Krishan,A. (1977) *Stain Technol.*, **52**, 339.
43. Stoehr,M., Hutter,K.J., Frank,M. and Goerttler,K. (1982) *Histochemistry*, **74**, 57.
44. Tien Kuo,M. (1982) *Exp. Cell Res.*, **138**, 221.
45. Collard,J.G., Tulp,A., Stegman,J., Boezeman,J., Bauer,F.W., Jonkind,J.F. and Verkerk,A. (1980) *Exp. Cell Res.*, **130**, 217.
46. Collard,J.G., Tulp,A., Stegeman,J. and Boezeman,J. (1981) *Exp. Cell Res.*, **133**, 341.
47. Van den Engh,G., Trask,B., Cram,S. and Bartholdi,M. (1984) *Cytometry*, **5**, 108.
48. Jensen,R.H., Langlois,R.G. and Mayall,B.H. (1977) *J. Histochem. Cytochem.*, **25**, 954.
49. Latt,S.A. and Wohlleb,J. (1975) *Chromosoma*, **52**, 297.
50. Behr,W., Honikel,K. and Hartmann,G. (1969) *Eur. J. Biochem.*, **9**, 82.
51. Le Pecq,J.B. and Paoletti,C.J. (1967) *Mol. Biol.*, **27**, 87.
52. Langlois,R.G., Carrano,A.V., Gray,J.W. and Van Dilla,M.A. (1980) *Chromosoma*, **77**, 229.

53. Latt,S.A., Sahar,E., Eisenhard,M.E. and Juergens,L.A. (1980) *Cytometry,* **1**, 2.
54. Cram,L.S., Arndt-Jovin,D.J., Grimwade,B.G. and Jovin,T. (1979) *J. Histochem. Cytochem.*, **27**, 445.
55. Gray,J.W., Langlois,R.G., Carrano,A.V., Burkhart-Schultz,K. and Van Dilla,M.A. (1979) *Chromosoma (Berl),* **73**, 9.
56. Gray,J.W., Peters,D., Merrill,J.T., Martin,R. and Van Dilla,M.A. (1979) *J. Histochem. Cytochem.*, **27**, 441.
57. Gray,J.W., Lucas,J., Pinkel,D., Peters,D., Ashwoth,L. and Van Dilla, M.A. (1980) in *Flow Cytometry IV,* Universitetsforlaget, Norway, p. 249.
58. Langlois,R.G., Yu,L.C., Gray,J.W. and Carrano,A.V. (1982) *Proc. Natl. Acad. Sci. USA,* **79**, 7876.
59. Bartholdi,M.F., Sinclair,D.C. and Cram,L.S. (1983) *Cytometry,* **3**, 295.
60. Wirchubsky,Z., Perlmann,C., Lindsten,J. and Klein,G. (1983) *Int. J. Cancer,* **32**, 147.
61. Cram,L.S., Bartholdi,M.F., Ray,F.A., Travis,G.L. and Kraemer,P.M. (1983) *Cancer Res.*, **43**, 4828.
62. Otto,F.J. and Oldiges,H. (1980) *Cytometry,* **1**, 13.

Suppliers of Specialist Items

Agfa Gevaert Ltd., French Ave., Dunstable, Beds, UK.

Aldrich Chemical Co. Ltd., The Old Brickyard, New Road, Gillingham, Dorset SP8 4JL, UK and 940W St. Paul Ave., P.O. Box 355, Milwaukee, WI 53201, USA.

Amersham International plc, UK Sales Office, Lincoln Place, Green End, Aylesbury, Bucks HP20 2TP, UK.

Assab Ltd., (See Don Whitley Scientific Ltd.)

BDH Chemicals Ltd., Broom Road, Poole, Dorset BH12 4NN, UK.

Becton Dickinson Ltd., Between Towns Road, Cowley, Oxford OX4 3LY, UK and Williams Drive, Oxnard, CA 93030, USA.

(Clay Adams Division) 299 Webro Road, Parsippany, NJ 07054, USA.

Bellco Glass Ltd., 340 Edrudo Road, Vineland, NJ 08360, USA.

Bethesda Research Laboratories, P.O. Box 6009, Gaithersburgh, MD 20877, USA.

Biological Industries Ltd., 56 Telford Road, Cumbernauld, Glasgow G67 2AX, UK.

Calbiochem-Behring Corp., La Jolla, CA 92037, USA.

Ciba-Corning Diagnostics Ltd., St. Andrews House, Halstead, Essex CO9 2DX, UK.

Chance Propper Ltd., P.O. Box 53, Spon Lane, Smethwick, Warley, Worcs. B66 1NZ, UK.

Clay Adams, (See Becton Dickinson.)

Costar, 205 Broadway, Cambridge, MA 02139, USA and Sloterweg 305-A 1171 VC Badhoevedorp, The Netherlands. (See Northumbria Biologicals Ltd., for UK distribution.)

Difco Laboratories Ltd., P.O. Box 14B, Central Avenue, East Molesey, Surrey KT8 OSE, UK and P.O. Box 1058, Detroit, MI 48232, USA.

Eastman Kodak Co., 343 State Street, Rochester, NY 14650, USA.

Fisher Scientific Co. Allied Corp., 711 Forbes Ave, Pittsburgh, PA 15219, USA.

Flow Laboratories Ltd., Woodcock Hill, Harefield Road, Rickmansworth, Herts. WD3 1PQ, UK and 7655 Old Springhouse Road, McLean, VA 22102, USA.

Gallenkamp, Belton Road West, Loughborough LE11 OTR, UK.

Gibco Ltd., Trident House, P.O. Box 35, Renfrew Road, Paisley PA3 4EF, UK and 3175 Staley Road, Grand Island, NY 14072, USA.

Grant Instruments (Cambridge) Ltd., Barrington, Cambridge CB2 5QZ, UK.

Hana Biologics Inc., 626 Bancroft Way, Berkeley, CA 94710, USA. (See Metachem Diagnostics Ltd. for UK distribution.)

Heico Wittaker, Philadelphia, USA.

Heraeus Christ GmbH, Gipsmuehlenweg 62, P.O. Box 1220, D-3360 Osterode, Harz, FRG.

Hoechst U.K. Ltd., Pharmaceutical Division, Hoechst House, Salisbury Road, Hounslow, Middlesex TW4 6JH, UK.

Arnold R. Horwell Ltd., 73 Maygrove Road, West Hampstead, London NW6 2BP, UK.

Hotpack/Heinicke Corp., 10940 Dutton Rd., Philadelphia, PA 19154, USA.

IBF Reactifs - Societe Chimique Pointet-Girard, 35 Avenue Jean-Jaures, F-92390 Villeneuve-la-Garenne, France.

ICN Biomedicals GmbH, Postfach 369, D-3440 Eschwege, FRG and P.O. Box 19536, Irvine, CA 92713, USA.

Ilford Ltd., Mobberley, Cheshire, UK.

Imperial Laboratories Ltd., Ashley Road, Salisbury, Wilts. SP2 7DD, UK.

Jencons (Scientific) Ltd., Cherrycourt Way Industrial Estate, Stanbridge Road, Leighton Buzzard, Beds. LU7 8UA, UK

Kodak Ltd., P.O. Box 66, Station Road, Hemel Hempstead, Herts. HP1 1JU, UK. (See also Eastman Kodak for USA address.)

Labco Ltd., 54 Marlow Bottom Road, Marlow, Bucks., UK.

Raymond A. Lamb, 6 Sunbeam Road, London NW10 6JL, UK.

Lederle Laboratories, Fareham Road, Gosport, Hants. PO13 OAS, UK.

Leec Ltd., Private Road No. 7, Colwick Industrial Estate, Nottingham NG4 2AJ, UK.

Leitz (Instruments) Ltd., 48 Park Street, Luton LU1 3HP, UK and 24 Link Drive, Rockleigh, NJ 07647, USA.

L.I.P. (Equipment and Services) Ltd., 111 Dockfield Road, Shipley, West Yorkshire BD17 7AS, UK.

LKB Instruments Ltd., Box 305, S-161 26 Bromma, Sweden, LKB House, 232 Addington Road, Selsdon, South Croydon, Surrey CR2 8YD, UK and 9319 Gaither Road, Gaithersburgh, MD 20877, USA.

Lux Scientific Corp., 1157 Tourmaline Drive, Newbury Park, CA 91320, USA.

Ma-re srl, I-20094 Corsico (Milano) - Via L. da Vinci, 47, Italy.

M.D.H., Walworth Road, Andover, Hants. SP10 5AA, UK.

Metachem Diagnostics Ltd., 29 Forest Road, Piddington, Northampton NN7 2DA, UK.

Millipore (UK) Ltd., Millipore House, 11-15 Peterborough Road, Harrow, Middlesex HA1 2YD, UK.

M.S.E. Scientific Instruments, Manor Royal, Crawley, West Sussex RH10 2QQ, UK.

Nalge Co. Nalgene Labware Dept., 75 Panorama Creek Drive, P.O. Box 365, Rochester, NY 14602, USA.

Nikon Inc., Instrument Group, 623 Stewart Ave., Garden City, NY 11530, USA and Building 72, P.O. Box 7609, 111725 Schiphol-Oost, The Netherlands.

Northumbria Biologicals Ltd., Unit 14, South Nelson Industrial Estate, Cramlington, Northumberland NE23 9HL, UK.

Nunc, A/S Nunc, Postbox 280, Kamstrup, FK 4000, Roskilde, Denmark. (See Gibco Ltd. for UK distribution.)

Olympus Corp. Scientific Instrument Div., 4 Nevada Dr., Lake Success, NY 11042, USA.

Reichert Scientific Instruments Div. of Warner Lambert Technologies Inc., P.O. Box 123, Buffalo, NY 14240, USA and C. Reichert Optische Werke AG, Henalser Hauptstrasse 219, A-1130 Wien, Austria.

Sera-lab Ltd., Crawley Down, Sussex RH10 4FF, UK.

Sigma Chemical Company Ltd., Fancy Road, Poole, Dorset BH17 7NH, UK and P.O. Box 14508, St. Louis, MO 63178, USA.

E. R. Squibb and Sons Ltd., Squibb House, Staines Rd., Hounslow TW3 3TA, UK.

Sterilin Ltd., Sterilin House, Clockhouse Lane, Feltham, Middlesex TW14 8QS, UK.

Sterling Organics, New York, USA.

Vindon Scientific Ltd., Ceramyl Works, Diggle, Oldham, Lancs OL3 5JY, UK.

Wacker Chemie, Munich, FRG.

Wellcome Diagnostics, Temple Hill, Dartford DA1 5AH, UK.

Whatman Labsales Ltd., Unit 1, Coldred Road, Parkwood, Maidstone, Kent ME15 9XN, UK.

Don Whitley Scientific Ltd., Green Lane, Baildon, Shipley, West Yorkshire BD17 5JS, UK.

Worthington Biochemical Corp., P.O. Box 650, Halls Mill Road, Freehold, NJ 07728, USA.

Carl Zeiss (Oberkochen) Ltd., P.O. Box 78, Woodfield Road, Welwyn Garden City, Herts AL7 1LU, UK and One Zeiss Dr., Thornwood, NY 10594, USA.

INDEX

Published
in the
Practical
Approach
series

Human genetic diseases

a practical approach

Edited by K E Davies, *University of Oxford*

Human genetic diseases

a practical approach

Edited by
K E Davies

Practical Approach Series

◯ IRL PRESS

June 1986; 152pp;
hardbound:
0 947946 76 4
softbound:
0 947946 75 6

This book gives diagnostic and research workers a complete guide to using the techniques of molecular biology for identifying genetic diseases in patients: it details the necessary methodologies from established and straightforward Southern blotting procedures to a unique account of the latest developments of pulsed-field gel electrophoresis. Molecular biology has brought many advances in clinical fields – in cancer research and identifying the AIDS virus, for example. Now the techniques in this book will assist in analysing specific mutations in patients and in isolating new markers for human genetic disease.

Contents

Fetal DNA analysis *J M Old* ● A short guide to linkage analysis *J Ott* ● The use of synthetic oligonucleotides as specific hybridization probes in the diagnosis of genetic disorders *S L Thein and R B Wallace* ● Alternative methods of gene diagnosis *J L Woodhead, R Fallon, H Figueiredo, J Langdale and A D B Malcolm* ● Fine mapping of genes, the characterization of the transcriptional unit *M Antoniou, E deBoer and F Grosveld* ● *In situ* hybridization *V J Buckle and I W Craig* ● Human chromosome analysis by flow cytometry *B D Young* ● Restriction analysis of chromosomal DNA in a size range up to two million base pairs by pulsed-field gradient electrophoresis *G J B van Ommen and J M H Verkerk*

For details of price and ordering consult our current catalogue or contact:

IRL Press Ltd,
PO Box 1, Eynsham,
Oxford OX8 1JJ, UK
IRL Press Inc,
PO Box Q,
McLean, VA 22101-0850,
USA

⬡ **IRL PRESS**
Oxford · Washington DC